Basic Farm Machinery

BY

J. M. SHIPPEN
Lincolnshire College of Agriculture and Horticulture, England

C. R. ELLIN
Welsh Agricultural College, Wales

C. H. CLOVER
Lincolnshire College of Agriculture and Horticulture, England

Third Edition

PERGAMON PRESS
OXFORD · NEW YORK · TORONTO · SYDNEY · FRANKFURT

U.K.	Pergamon Press Ltd., Headington Hill Hall, Oxford OX3 0BW, England
U.S.A.	Pergamon Press Inc., Maxwell House, Fairview Park, Elmsford, New York 10523, U.S.A.
CANADA	Pergamon of Canada, Suite 104, 150 Consumers Road, Willowdale, Ontario M2J 1P9, Canada
AUSTRALIA	Pergamon Press (Aust.) Pty. Ltd., P.O. Box 544, Potts Point, N.S.W. 2011, Australia
FEDERAL REPUBLIC OF GERMANY	Pergamon Press GmbH, Hammerweg 6, D-6242 Kronberg-Taunus, Federal Republic of Germany

First edition (in two volumes) 1966

Second edition 1973

Reprinted 1974, 1975, 1976, 1978 (twice)

Third edition 1980

Reprinted (with corrections) 1984, 1985

British Library Cataloguing in Publication Data
Shippen, John Matthew
Basic farm machinery. - 3rd ed.
- (Pergamon international library).
1. Agricultural machinery
I. Title   II. Ellin, C. R.   III. Clover, C. H.
631.3   S675   79-41103

ISBN 0-08-024912-4 (Hardcover)
ISBN 0-08-024911-6 (Flexicover)

Printed in Great Britain by A. Wheaton & Co., Ltd., Exeter

Contents

Preface to the Third Edition

THE original volume of this book was published in 1966 and in the Preface it was stated that its purpose was to deal with the general working principles of farm tractors and machinery. This purpose remains so but of course, because of progress within the industry, development and introduction of new machines and methods, it has been found necessary again to revise and add to the contents.

The subject matter of the book is very suitable for the agricultural student, the agricultural apprentice, and the apprentice agricultural mechanic. Particular attention has been paid to the needs of the student studying for City and Guilds of London Institute examinations in Tractors, Farm Implements and Machinery at Phase 1 and 2 levels and for similar examinations of the regional examining bodies. Many parts will also be relevant to the needs of horticultural students.

The book should also be of value to agricultural and horticultural workers who have the necessary interest in their work. It should help them to understand the general working principles of the common types of machinery, and thus enable them to profit fully from any further instruction given at educational or training courses on the detail of any particular make or type of machine.

It will be noted that irrelevant detail has been omitted from the drawings and line diagrams which illustrate the text in order that the student should get a clear idea of the working principles involved. A further aid to full understanding is the use of several diagrams to illustrate different mechanisms or phases of a cycle on the same machine.

At the end of each chapter is also given information on servicing and maintenance to the farm machines discussed. This is a very important aspect of machinery care and use.

Also included is a chapter on farm safety which is all important in this age of mechanization on farms. Many regulations now cover the guarding and use of farm machines and tractors and all machinery operators should have some knowledge of these and of the risks involved when handling machinery.

Although the majority of machines illustrated are common in the United Kingdom and Europe, many of the same machines are used in other parts of the world; furthermore, apart from a few examples of very specialized machines, the same general principles apply to all the main types of agricultural equipment. Consequently, a great deal of the subject matter in this book should be of value to students and agricultural workers anywhere in the world.

It will be noted also that both British and metric units of measure have been used. The phased change over from British Standard units of measure to the metric measure is taking place gradually and is not yet complete. Many people within the industry will continue to use British Standard measure for a long time to come, hence the need for the use of both within this book. The authors have made use of both units of measure by adding the metric equivalent of all British Standard units used in this book. By doing this, however, it will be seen that some rather odd metric measurements result, e.g. a wheel size of 12 × 36 in. becomes 304 × 914 mm or a 28-in. row spacing becomes a 711-mm row spacing. Of necessity and as a result of the conversion factors involved, some of the metric equivalents have been made approximations, e.g. a 12-in.-wide wheel is in fact 304·8 mm, but in such cases the decimal fractions of millimetres have been ignored. Throughout the book no single unit of metric measure of length has been used as is sometimes recommended. Millimetres, centimetres and metres have been used where considered appropriate.

A list of conversion factors is provided in the book which should be of use to all readers.

Preface to the Second Edition

THE main purpose of the Second Edition of this book has changed in no way from that of the First Edition. It has been found necessary to add to the contents but the additions are limited on the grounds of size and cost; however, such additions and revision that have taken place have widened the scope of the book.

The purpose of the original two volume book was to deal with the general working principles of farm tractors and machinery. This is still so, and the types of reader who should find the subject matter of value include the agricultural student, the agricultural apprentice and the young agricultural mechanic who may be relatively expert on a narrow range of machinery but who should also have a general picture of the equipment used on farms.

Particular attention has been paid to the needs of the student sitting for Farm Machinery (Introductory) and Farm Machinery and Equipment examinations of the City and Guilds of London Institute and for the similar examinations of the regional examining bodies. Many parts will also be relevant to the needs of the horticultural student, particularly anyone sitting for the Stage I Horticultural Machinery examination of the City and Guilds.

The book should also be of value to agricultural and horticultural workers who are interested in their work; it should help them to understand the general working principles of the common types of machinery and thus enable them to profit fully from any further instruction given at educational or training courses on the detail of any particular make or type of machine.

It will be noted that irrelevant detail has been omitted from the drawings and line diagrams which illustrate the text in order that the student should get a clear idea of the working principles involved. A further aid to full understanding is the use of several diagrams to illustrate different mechanisms or phases of a cycle on the same machine.

At the end of each chapter is also given information on servicing and maintenance to the farm machines discussed. This is a very important aspect of machinery care and use.

Also included is a chapter on farm safety which is all-important in this age of mechanization on farms. Many regulations now cover the guarding and use of farm machines and tractors and all machinery operators should have some knowledge of these and of the risks involved when handling machinery.

Although the majority of machines illustrated are common in the United Kingdom and Europe, many of the same machines are used in other parts of the world; furthermore, apart from a few examples of very specialized machines, the same general principles apply to all the main types of agricultural equipment. Consequently, a great deal of the subject matter in this book should be of value to students anywhere in the world.

It will also be noted that both British and metric units of measure have been used. At the time of writing, the phased changeover to take place in the United Kingdom, from the British Standard units of measure to the metric units of measure, is not yet complete. In view of this, and at this point in time, there being no recommendations available regarding the S.I. units to be used in the agricultural industry, the authors have added the Metric equivalent of all British Standard units used in the book. By doing this, however, it will be seen that some rather odd metric measurements result, e.g. a wheel size of 12 × 36 in. becomes 304 × 914 mm or a 28-in. row spacing becomes a 711-mm row spacing. Of necessity and as a result of the conversion factors involved, some of the metric equivalents have been made approximations, e.g. a 12-in.-wide wheel is in fact 304·8 mm, but in such cases the decimal fractions of millimetres have been ignored. Throughout the book no single unit of metric measure of length has been used as is sometimes recommended. Millimetres, centimetres and metres have been used where considered appropriate.

The introduction of metric units of measure will inevitably in due course affect farm tractors and implements. It could well be that tractor wheel settings will in future be adjustable in steps of 5 cm instead of 2 in. However, as no positive information is yet available about this, this is merely thinking aloud.

A list of conversion factors is provided in the book which should be of use to all readers.

Acknowledgements

The authors are grateful to R. Ellin, N.D.Agric.E., for his help in revising the chapter on Electricity and also to Miss Brenda Johnson and Mrs. Mary Goodliffe for their excellent typing of the manuscript.

CHAPTER 1

The Agricultural Tractor

General

The importance of agricultural tractors in the world today should never be underestimated. They provide a great source of power which has almost entirely replaced the power of horses, and of man, for the many heavy and time-consuming tasks carried out on the land.

This power is being used to produce food for the nations of the world and there is no more important task than this. A tractor can do the work of numerous horses, and do this work in the same time, without having to be rested to recover from fatigue. Provided reasonable attention is given to its lubrication and it is kept supplied with fuels, it will work on indefinitely. The driver, of course, will require his rest, but it is not unusual for tractors to work around the clock when relief drivers are available.

As more and more food production is required to feed the ever-increasing world population, the numbers of tractors used will increase to meet the demand. Great areas of land will yet be brought into food production and the tractor will supply the necessary power for the work.

Tractor Types

The present-day tractor is a most useful machine, capable of supplying its power to numerous farm tasks. The most power absorbing of these tasks is usually the basic cultivation of land; therefore the tractor is designed to be able to do this task whilst at the same time being amply powered for the many other field tasks it is required to do, such as drilling seed, top dressing, spraying, haulage, etc.

During the years since the First World War, and in particular during, say, the last 20 years, a great amount of progress has been made in developing a machine capable of operating efficiently a very wide range of implements and machines. Many devices have been incorporated in the mechanism of the tractor for this purpose.

The type of tractor used on the land depends on the type of work to be done.

Track-laying Tractors (Fig. 1)

These tractors are sometimes referred to as "crawlers" and are usually tractors with a large horse-power and capable of doing very heavy work. On farms, they may be used for pulling a five- or six-furrow plough or for heavy cultivation. Yet, on the other hand, there is a very small track-laying tractor of no more than 6 h.p. (horsepower) which is used for work on market gardens.

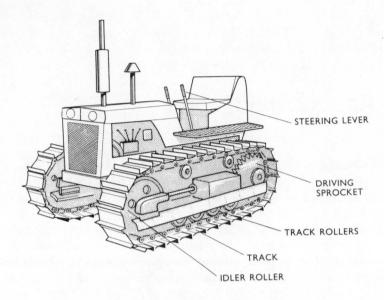

FIG. 1. A track-laying tractor.

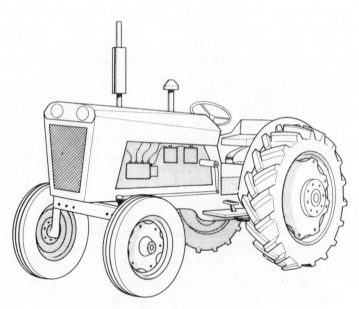

FIG 2. A general-purpose tractor.

Heavy Wheeled Tractors

Not only track-laying tractors are used for the heavy work on farms. Very large four-wheel-drive tractors are now common which are capable of pulling up to twelve furrow ploughs with engines of up to 500 h.p. These large tractors may be equipped with twin wheels all round, and no longer run in the furrow when ploughing, but run on top in the same way as a crawler tractor. They also carry out very heavy cultivations.

General-purpose Tractors (Fig. 2)

The general-purpose tractor is a wheeled tractor and the lighter type may be powered by an engine of about 40 h.p. whilst the other type may have an engine of 100 h.p. Either one or both of these types may be used on most farms and they are the most common type used today.

Two-wheeled Tractors

Another type of tractor to be found on many small holdings and market gardens is the hand-operated. This type is powerful enough to do light cultivations and is steered by a walking operator.

Most tractors nowadays are powered by internal combustion engines, which, although they may vary slightly between one make and another as regards detail of construction, operate on the same basic principles.

The Internal Combustion Engine

The tractor's power is used for doing work in the field and also for driving stationary machines.

The power, which is produced by the engine, is transmitted through various mechanisms until it reaches the rear wheels which rotate. These mechanisms will be explained in later chapters. The power unit must be considered first.

In any internal combustion engine there are a number of essential parts which are so arranged that a basic series of events occur. This series of events is usually known as the operating cycle and most tractor engines operate on what is known as the four-stroke cycle.

The internal combustion engine is a form of heat engine and the name "heat engine" is given to it because heat energy, produced by the burning of fuel within the engine, is changed into mechanical energy. Different types of fuel are used in the different types of tractor engines but this does not alter the basic operation. The fuels used are petrol or propane for spark ignition, and diesel fuel is used in compression ignition engines.

The Essential Components (Fig. 3)

The essential parts of this type of engine are as follows:
(a) A *cylinder* into which the fuel is introduced for burning.
(b) A *piston* which moves up and down within the cylinder.
(c) A *cylinder head* which seals off the top of the cylinder.
(d) A crankshaft which is rotated by the piston on power stroke.
(e) A connecting rod which connects the piston to the crankshaft. These two

components convert reciprocating motion of the piston into rotary motion of the crankshaft to form a useable means of power.

(f) A gudgeon pin which connects the connecting rod to the piston.

(g) An inlet valve which allows the fresh charge of fuel/air into the combustion chamber.

(h) An exhaust valve which allows the spent gases to escape to the atmosphere via the exhaust system.

(i) A camshaft which opens the valves.

(j) Valve springs which close the valves.

(k) A flywheel fitted to the crankshaft which absorbs power on the power stroke and gives off energy on the three non-power strokes to maintain a smooth operation of the engine.

Figure 3 shows these essential parts of the four-stroke engine and Fig. 4 shows a typical arrangement of these parts within the engine. Although the arrangement shown is for a single cylinder *vertical* engine, some engines have the pistons moving *horizontally*. Where an engine has a number of cylinders, the arrangement shown is simply multiplied,

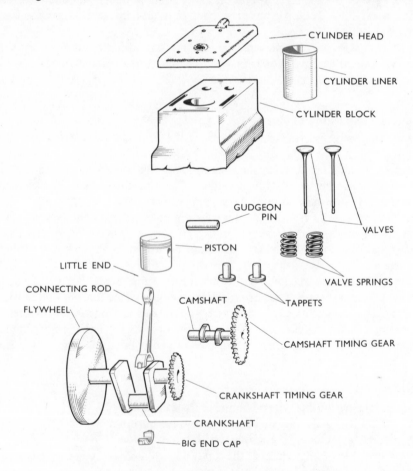

Fig. 3. Some of the main working parts of a side valve water-cooled single-cylinder engine.

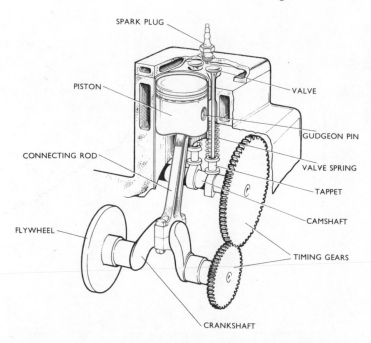

Fig. 4. A section through a single-cylinder engine showing the relative
position of the working parts.

so that all connecting rods would be attached to one crankshaft having the required
number of "cranks" along its length (see Fig. 5 for various layouts).

Tractor engines are usually four-cylinder engines but it is not uncommon for them
to have six cylinders; one, two and three cylinders are also used. Figure 6 shows a
crankshaft with pistons and connecting rods for a four-cylinder engine.

Whether an engine is vertical, horizontal, or is single- or multi-cylinder, does not
alter the method in which it operates. The basic cycle still takes place independently in
each cylinder of the engine; which means that in a multi-cylinder engine, each cylinder
with its piston, connecting rod and valves, etc., should be considered as an independent
mechanical unit.

From Fig. 4 it will be seen that if the flywheel is turned, the crankshaft will also
turn and this will cause the piston, which is free to move in the cylinder, to travel up
and down. Because the gear wheel on the crankshaft is in mesh with the gear on the
camshaft, the valves will also move up and down. If a thrust is applied to the top of the
piston, the flywheel and crankshaft will again turn and in doing so will also operate
the camshaft and valves.

This in fact is just what happens in the engine. Fuel/air mixture comes into the
cylinder when the inlet valve is open, and when the valve closes, the mixture is compressed
by the piston moving up. The mixture is then ignited and expands rapidly, and the
expansion produces a downward thrust to the piston which in turn causes the crankshaft
to rotate. The thrust is sufficient to keep the crankshaft turning so that the piston also
continues to move up and down. The movement of these parts in relation to each
other makes possible the operating cycle.

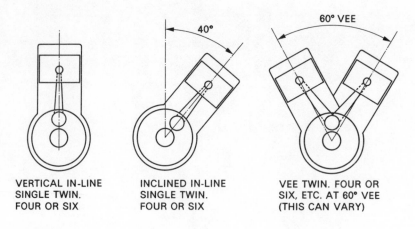

VERTICAL IN-LINE
SINGLE TWIN.
FOUR OR SIX

INCLINED IN-LINE
SINGLE TWIN.
FOUR OR SIX

VEE TWIN. FOUR OR
SIX, ETC. AT 60° VEE
(THIS CAN VARY)

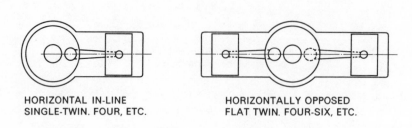

HORIZONTAL IN-LINE
SINGLE-TWIN. FOUR, ETC.

HORIZONTALLY OPPOSED
FLAT TWIN. FOUR-SIX, ETC.

Fig. 5. Engine layouts and terminology.

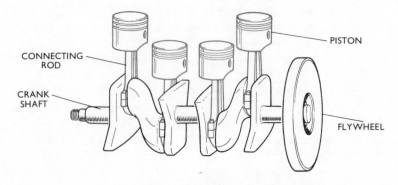

Fig. 6. A four-cylinder engine crankshaft with connecting rods
and pistons.

The Four-stroke Cycle (Spark Ignition)

Engines that are designed to run on petrol or propane require an electric spark to ignite the fuel/air mixture. These are known as Spark Ignition engines.

No. 1. The induction stroke. The piston moves down, inlet valve open, exhaust valve closed. This downward movement of the piston creates a partial vacuum inside the cylinder causing fuel/air mixture to rush in through the inlet valve passage. When the piston reaches the bottom of its stroke the inlet valve closes to prevent any escape of the fuel, therefore the fuel is trapped inside the cylinder.

No. 2. The compression stroke. The piston moves up, the inlet valve is closed, the exhaust valve is closed. This upward movement compresses the fuel/air mixture into a small space at the top of the cylinder. This space is usually provided by having the underside of the cylinder head suitably shaped, e.g. slightly concave.

No. 3. The power stroke. When the piston is at the top of the compression stroke with both valves closed, a spark is arranged to take place across the points of a spark plug which protrudes partly inside the combustion space. This spark immediately sets fire to the mixture which burns and expands, and in doing so forces down the piston; this provides the power.

No. 4. The exhaust stroke. The piston moves up the cylinder again, the inlet valve is closed, the exhaust valve is open. This upward movement pushes the burnt gases out through the exhaust valve outlet and to the atmosphere.

Figure 7 shows in diagram form the four strokes which make up the cycle of operation. The position of the valves as shown in the diagram is for the convenience of the reader. In practice they would be positioned side by side.

At the end of the exhaust stroke, the cycle starts again with a fresh charge of fuel/air mixture being taken into the cylinder.

Of the three strokes in the operating cycle, only one does the work. This is the third stroke which is the *power stroke.* The other three strokes make the power stroke possible, and the thrust of the power stroke keeps the crankshaft turning so that these strokes can take place. A stroke is one movement from top dead centre to bottom dead centre or vice versa.

The Two-stroke Engine (Spark Ignition)

A type of engine known as a two-stroke engine is often used to provide power for small machines used on farms or in horticultural work. Some makes of tractors are powered by these engines but the majority of modern tractors use the four-stroke engine. Typical examples of machines that may be powered by the two-stroke engine are hand-operated hedge trimmers, hand-operated chain saws, small rotary cultivators, rotary grass cutters and horticultural sprayers.

The construction of the engine is different to that of a four-stroke engine in that valves are not normally fitted. Instead, openings known as *ports* are arranged in the walls of the cylinder. These openings or ports are known as:

Inlet port. Where the carburettor is attached and fuel/air enters the crankcase.

Transfer ports (two). Where compressed fuel/air is transferred from the crankcase to the combustion chamber, ready for the compression stroke.

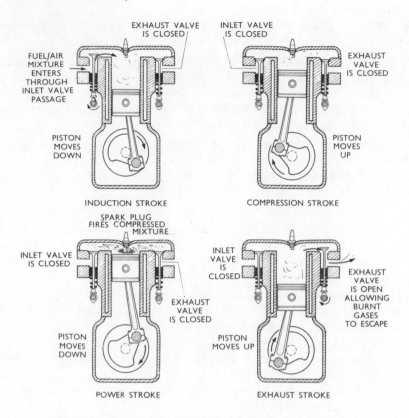

Fig. 7. The four-stroke cycle (spark ignition) side valve arrangement.

Exhaust port. At the end of the power stroke the piston uncovers the exhaust port and spent gases escape.

The crankcase is a sealed compartment and acts as a fuel/air pump for the engine. Unlike the four-stroke cycle, the two-stroke engine sends the piston down on every downward stroke. Reference to Figs. 8a and b will help the reader to understand how the cycle operates.

The situation shown in Fig. 8a is that the piston is almost at the top of the compression stroke compressing fuel mixture above it, and from this point the operating cycle can be followed.

The Two-stroke Cycle

Stroke 1. As the piston moves up on compression of a previous charge, a partial vacuum is created in the sealed crankcase.

The piston uncovers the inlet port and air rushes through the carburettor taking fuel/oil mixture with it into the crankcase.

Stroke 2. The previous charge now ignites, pushing the piston down on power stroke, this seals the fuel/air in the crankcase and compresses it. When the piston reaches bottom dead centre the exhaust ports are opened and the spent gases escape.

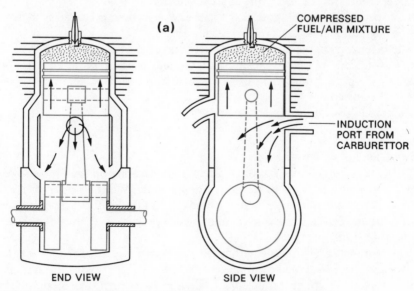

(a)

COMPRESSED
FUEL/AIR MIXTURE

INDUCTION
PORT FROM
CARBURETTOR

END VIEW

SIDE VIEW

INDUCTION AND COMPRESSION STROKE SHOWING CRANKCASE INDUCTION

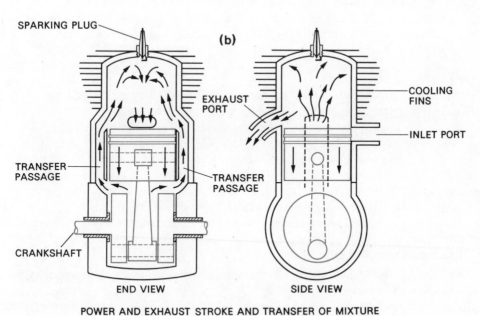

SPARKING PLUG

(b)

EXHAUST
PORT

COOLING
FINS

INLET PORT

TRANSFER
PASSAGE

TRANSFER
PASSAGE

CRANKSHAFT

END VIEW

SIDE VIEW

POWER AND EXHAUST STROKE AND TRANSFER OF MIXTURE

Fig. 8. Schnurle type two-stroke engine.

Also the transfer ports are opened and the compressed fuel/air mixture enters the cylinder from both transfer ports at great velocity.

The special design of these ports causes the incoming charge to deflect upwards thereby avoiding the exhaust port which is open.

The incoming charge also aids complete scavenge of used exhaust gases (in theory). The piston now moves up closing transfer and exhaust ports and compressing the charge. A spark ignites the mixture just before top dead centre and power stroke occurs thereby completing the cycle.

Refer to Fig. 8 for the cycle of operation.

The Purpose of a Flywheel

To make an engine run smoothly when it is operating on the four-stroke cycle, it is necessary to fit a heavy flywheel to the crankshaft. When a power stroke takes place once in every four strokes, the engine tends to vibrate. A flywheel gathers momentum as it is turning and this assists in the turning of the crankshaft when the three non-power strokes are taking place.

The more cylinders an engine has, the smoother it can be made to run. Figure 6 shows four pistons connected to one crankshaft and they are arranged to move up and down in pairs. With an arrangement like this in an engine, it is possible to have the power stroke of each of the cylinders taking place at a different time. When the first cylinder is on the power stroke, the second could be on exhaust, the third on compression and the fourth on the inlet. This arrangement gives us what is known as a *firing order,* this being the order in which each of the cylinders is on the power stroke. A typical firing order is 1, 3, 4, 2. With this sort of arrangement, a lighter weight flywheel may be used because a power thrust is applied to each of the four pistons at a different time.

The Diesel Engine

This engine is made very much the same as the petrol or vaporizing oil engine, but because higher pressure and thrusts take place within the engine itself, it is made stronger. The basic parts are very much the same as in the other types of engines, the method of operation is slightly different. Instead of fuel/air mixture being taken into the cylinder, air alone enters and this is compressed to such an extent that it becomes very hot. It becomes so hot in fact that it will set fire to fuel injected into it. These engines can be designed to operate on the four-stroke cycle or two-stroke cycle and in the past some tractor diesel engines have operated on the two-stroke cycle.

The diesel engine is also sometimes known as a compression ignition engine and no electric spark is required to ignite the fuel.

The operating cycle is as follows:

No. 1. The induction stroke. The piston moves down, inlet valve open, exhaust valve closed. The downward movement of the piston creates a partial vacuum in the cylinder and air only rushes in through the inlet valve passage. At the bottom of the stroke the inlet valve closes trapping the air within the cylinder.

No. 2. The compression stroke. The piston moves up, the inlet valve is closed, the exhaust valve is closed. The upward movement of the piston compresses the air to a very high temperature, about 1000°F. This temperature is reached because the volume

TWO-CYLINDER ENGINE

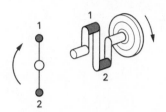

FIRING ORDER 1-2-1-2
360° CRANKSHAFT ROTATION
BETWEEN EACH POWER STROKE

FOUR-CYLINDER ENGINE

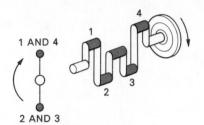

FIRING ORDER 1-3-4-2 OR
        "   *"*   1-2-4-3
180° CRANKSHAFT ROTATION
BETWEEN EACH POWER STROKE

THREE-CYLINDER ENGINE

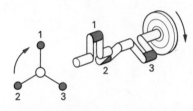

FIRING ORDER 1-2-3
240° CRANKSHAFT ROTATION
BETWEEN EACH POWER STROKE

SIX-CYLINDER ENGINE

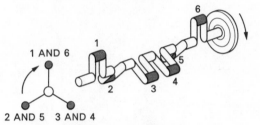

FIRING ORDER 1-5-3-6-2-4 OR
        "   *"*   1-4-2-6-3-5
120° CRANKSHAFT ROTATION
BETWEEN EACH POWER STROKE

FOR TWO-STROKE ENGINES THE ABOVE DEGREES OF CRANKSHAFT ROTATION WOULD BE HALVED

Fig. 9. Multi-cylinder engine firing orders.

within the cylinder is reduced to approximately 16 times its original volume. In a spark ignition engine it would be reduced about 7 or 8 times.

No. 3. The power stroke. When the piston is at the top of the compression stroke, a spray of fuel is injected into the cylinder. This fuel ignites immediately when it comes in contact with the hot air. The burning fuel/air mixture expands and thrusts the piston down the cylinder.

No. 4. The exhaust stroke. The piston moves up the cylinder, the inlet valve is closed, the exhaust valve is open. The upward movement of the piston pushes the burnt gases out through the exhaust valve passage and to the atmosphere.

Figure 10 shows in diagram form the four strokes of the diesel engine operating cycle.

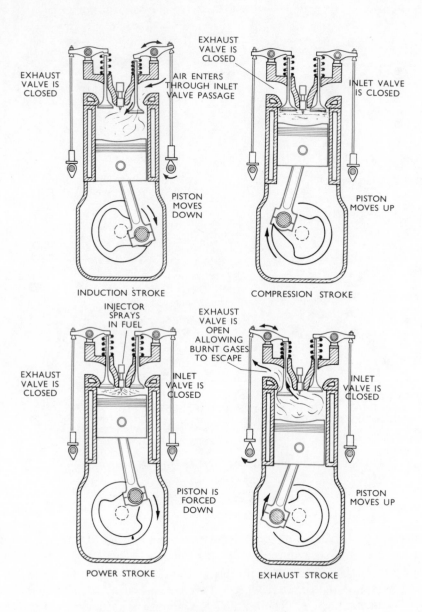

EXHAUST
VALVE IS
CLOSED

EXHAUST
VALVE IS
CLOSED

AIR ENTERS
THROUGH INLET
VALVE PASSAGE

PISTON
MOVES
DOWN

INDUCTION STROKE

INLET VALVE
IS CLOSED

PISTON
MOVES UP

COMPRESSION STROKE

INJECTOR
SPRAYS
IN FUEL

EXHAUST
VALVE IS
CLOSED

INLET
VALVE IS
CLOSED

PISTON IS
FORCED
DOWN

POWER STROKE

EXHAUST
VALVE IS
OPEN
ALLOWING
BURNT GASES
TO ESCAPE

INLET
VALVE IS
CLOSED

PISTON
MOVES UP

EXHAUST STROKE

Fig. 10. The four-stroke cycle (compression ignition engine).

Valve Arrangements

All tractor engines have overhead valve design as this is the only suitable arrangement to achieve the high compression ratios needed for compression ignition engines. The side-valve arrangement is still used on many smaller engines for driving stationary farm machines.

(a) DIRECT INJECTION COMBUSTION CHAMBER

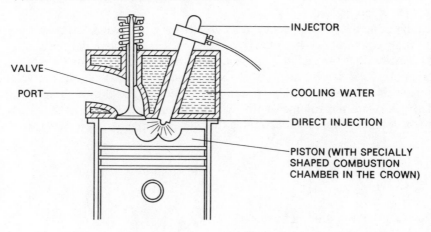

INJECTOR

VALVE

PORT

COOLING WATER

DIRECT INJECTION

PISTON (WITH SPECIALLY SHAPED COMBUSTION CHAMBER IN THE CROWN)

(b) INDIRECT INJECTION COMBUSTION CHAMBER

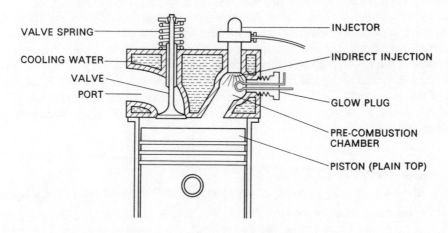

VALVE SPRING

COOLING WATER

VALVE

PORT

INJECTOR

INDIRECT INJECTION

GLOW PLUG

PRE-COMBUSTION CHAMBER

PISTON (PLAIN TOP)

Fig. 11. Diesel engine combustion chamber design.

Compression Ratios

When the piston is at the bottom of the induction stroke there is a certain volume in the cylinder that is filled with fuel mixture or air. When the piston moves up on the compression stroke, this volume of mixture or air is compressed into the small space at the top of the cylinder. The space is known as the *combustion chamber.*

To find the *compression ratio,* all we have to do is to work out the ratio between the volume in the cylinder to the volume in the combustion space when the piston is at the bottom of the stroke. By doing this we will know the extent to which the fuel mixture is compressed.

Supposing the volume of a cylinder with the piston down is 30 in³, and when it is up the volume of the space at the top is 5 in³ then the compression ratio will be 30 ÷ 5 = 6:1. This means that the mixture has been compressed into a volume 6 times less than the original. Or it can be calculated mathematically by working out the volume of the combustion chamber when the piston is at T.D.C. to give the clearance volume (C.V.). Then calculate the volume of the cylinder when the piston is at B.D.C. to give the swept volume (S.V.). By substituting the calculated figures in the formula

$$\frac{\text{C.V.} + \text{S.V.}}{\text{C.V.}} = \text{C.R. (compression ratio)}$$

the compression ratio will be found.

The compression ratio of an engine is an important feature because it is related to the quality and type of fuel that can be used, and it has a direct part in deciding the power output and efficiency of any internal combustion engine. Petrol engines for farm tractors have a low compression ratio of about 5·5 to 1 up to 7 to 1. Some small horticultural or farm stationary engines can be as low as 5 to 1. This enables them to run quite happily on low octane fuel which is ignited by an electric spark.

Diesel engines have a much higher compression ratio as they rely on the air being compressed and therefore heated, to ignite the fuel which is injected into the combustion chamber. Minimum compression ratios for diesel engines are 15 to 1, but most modern diesels now operate with a 20 to 1 compression ratio. The air when fully and speedily compressed can reach temperatures of 1112°F (600°C), which is well above the flash point of the atomized diesel fuel. To attain this combustion heat the engine must be turned over very quickly by the starter or starting may be difficult (a minimum of 200 r.p.m. is usually required).

Diesel Engine Starting Aids

For cold-climate starting of diesel engines various methods are used to aid starting:

1. *Glow-plugs.* These are located in the pre-combustion chambers, one plug per cylinder. These glow-plugs are heated by the tractor batteries and glow continuously while the switch is being operated.

A ballast resistor indicator is usually fitted on the dash panel to warn the operator that the plugs are glowing.

Immediately the indicator glows the starter should be operated and starting is usually achieved quickly if everything is in good order.

2. *Thermostart.* This is a single heater unit fitted in the inlet manifold to heat the ingoing air. A key type switch operates the heater element; 15-20 seconds is the usual amount of time needed to glow the element, no indicator is fitted.

The glowing element activates a bi-metal strip which then uncovers an orifice to allow fuel to flow on to the glowing element, this sets fire and can be heard in the inlet manifold. On hearing this noise immediately operate the starter, drawing the flame into the engine which should give immediate starting.

3. *Excess-fuel device.* On some tractor fuel pumps there is a small button to depress or a small T-bar to turn clockwise which allows extra fuel to spray into the engine to aid starting. As the engine starts the small button flies out automatically to reduce the amount of fuel, but the T-bar type must be released manually or the engine will run very erratically and rich.

4. *Ether start.* An ether-based spray is available in aerosol cans which aids starting by spraying a very small amount into the inlet manifold of the engine as the starter is turning the engine over. This gives instant starting to virtually any engine, however poor the engine's condition may be. Before using this starting method a few simple rules must be read on the can or the engine may be permanently damaged, or worse still human injury may occur.

The main points to observe are:

(a) The spray is highly flammable, therefore do not spray near naked flame, such as cigarettes.

(b) The engine must be turning over on the starter with no heaters in use before spraying into the air intake. Only a very short spray is required. Prolonged spraying into the intake can cause severe combustion explosions and detonations which can fracture crank, con-rods or cylinder heads.

Glow plug starting aid. (One plug per cylinder). The plugs are wired in "series". Therefore if one burns out, the rest fail to operate as an engine protection measure.

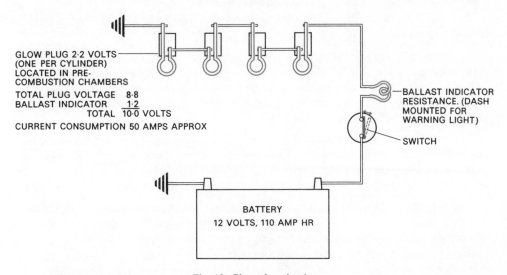

GLOW PLUG 2·2 VOLTS
(ONE PER CYLINDER)
LOCATED IN PRE-
COMBUSTION CHAMBERS
TOTAL PLUG VOLTAGE   8·8
BALLAST INDICATOR   1·2
              TOTAL 10·0 VOLTS
CURRENT CONSUMPTION 50 AMPS APPROX

BALLAST INDICATOR
RESISTANCE. (DASH
MOUNTED FOR
WARNING LIGHT)

SWITCH

BATTERY
12 VOLTS, 110 AMP HR

Fig. 12. Glow plug circuit.

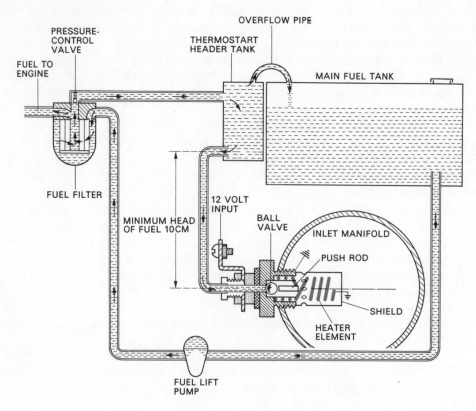

Fig. 13. Thermostart starting aid showing fuel feed circuit.

Four-stroke Diesel Engine Combustion Chamber Designs

Two main types of diesel engine combustion chamber designs are employed on tractor engines:

(a) *The direct type.* This is where the injector is fitted to spray directly on to the piston crown, which may be shaped to form a combustion chamber. Usually a multi-hole injector is used which spreads the spray evenly over the chamber to give good fuel air mixing. The type of engine usually starts easily and quickly without any special aids, such as heaters or excess fuel devices.

(b) *The indirect type.* This is where the injector sprays through a single-hole or double-hole injector at slightly lower pressure into a pre-combustion chamber coupled to the main chamber by a tangential passage. Ignition or burning first occurs in the pre-combustion chamber then passes through into the main chamber.

This type of engine very seldom starts without the aid of heaters in cold weather. Although this type of engine has the disadvantage of needing heaters to start it, it has certain advantages—its usually more economical on fuel, has a better thermal efficiency and will pull heavy loads without a fall-off in revs like the direct type, also a lower injector pressure increases the life of the injectors and fuel pump. See Fig. 11 for sectioned view of direct and indirect combustion chamber design.

CHAPTER 2

The Cooling System

THE internal combustion engine burns its fuel inside the cylinder to produce power. The heat produced on the power stroke of a diesel engine can be as high as 1600°C and this is greater than the melting point of the engines parts that come in contact with the heat. For example, melting point of cast iron is 1100°C, of steel 1450°C and of aluminium alloy 660°C. The cylinder and cylinder head are usually cast iron and the piston aluminium alloy, thus it is now easy to understand why an engine must have an efficient cooling system.

Assuming the heat value of the fuel used to be 100% then probably 30% is used to produce power, 40% is passed to the atmosphere via the exhaust and 30% is dissipated by the cooling system. Some heat is also removed by the oil employed in the lubrication system.

It is very important that the engine does not overheat or "boil the water", but it is equally important that the engine does not run too cool or the fuel will not vaporize properly, again causing engine damage and rapid wear.

Air Cooling

This method is used mainly for small single-cylinder engines, such as those fitted to horticultural machines, or which may be found around the farm driving such things as elevators or potato sorters.

Cooling in this case is done by directing a blast of air around the cylinder and cylinder head, both of which are finned. This finning provides a large surface area through which the heat can spread and from which heat is taken away by the cooling air.

The blast of air is created by fitting a blower fan to the engine flywheel and this is encased in a cowling which extends over and around the cylinder. Air is drawn in by the blower and directed over the cylinder to cool it. Figure 14 shows a typical air cooling arrangement.

Overheating of this type of engine will take place if the spaces between the fins are allowed to become choked with dirt, because the surface area is reduced.

Water Cooling

The majority of tractor engines are water cooled. This is a more effective method than air cooling for a multi-cylinder engine.

The engine cylinders are completely surrounded by a water jacket which also extends into the cylinder head, and from the cylinder head an outlet is connected by a rubber hose to the top of the radiator. From another outlet at the bottom of the radiator, a hose connection is taken to the water jacket surrounding the cylinder. This provides a path through which the water can flow.

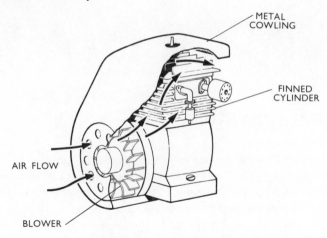

Fig. 14. An air-cooled engine.

A radiator is constructed to provide the largest possible area of surface that can be exposed to fan assisted air flow. A radiator is made up of a header tank and a bottom tank which are connected together by a lot of thin tubes through which the water flows (Fig. 15). The tubes are also in contact with numerous thin metal plates and this provides a further increase in surface area. The whole radiator looks like a honeycomb through which air can flow.

Behind the radiator and in front of the engine is a fan which is driven by a pulley on the engine crankshaft.

Water Circulation

As soon as the engine is started, heat is produced in the cylinders and this heat is transferred to the water surrounding them. When water is heated it becomes less dense, in other words warm water is lighter than cold, and because of this the heated

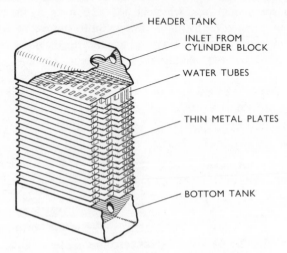

Fig. 15. A section through a tractor radiator.

water rises up into the radiator header tank. It is then replaced by cool water from the bottom of the radiator. This natural movement of the water is known as *thermo-syphon*.

As the heated water flows into the radiator tubes its heat is given off to the atmosphere and this is helped by the flow of cool air drawn through the honeycombed radiator by the fan.

The Impeller

Thermo-syphon circulation by itself is not satisfactory for the modern engine, therefore movement of the water is usually helped by an *impeller*. This impeller is simply a metal disc with vanes around one of its faces. It is positioned in the cooling system and driven by the fan shaft. As it rotates it draws cooler water up from the bottom of the radiator, thus helping circulation.

The Thermostat

An additional device usually fitted into the cooling system is a *thermostat*. This is used to help control the temperature of the water in the vicinity of the cylinders. It is very useful when an engine is started up from cold or when it has to "idle" (run slowly) for a long time.

The thermostat is a form of automatic valve fitted into the outlet pipe connecting the cylinder head to the top of the radiator. It restricts the flow of water to the radiator until an efficient working temperature is reached within the cylinder block.

The flow of water is either stopped, has limited movement or unrestricted movement according to the temperature of the water. Because the volume of water within the cylinder block is much less than that within the whole cooling system, it means that rapid warming up from cold takes place. The water in the block is circulated by the water pump while the thermostat is closed, a by-pass is incorporated in the circuit to allow this internal circulation. On some engines the thermostat opens and closes this by-pass circuit. This internal circulation prevents localised hot spots forming which may damage parts of the engine castings. Figure 16 shows the layout of a typical cooling system in a modern engine.

A suitable working temperature of an engine is about 90°C (194°F) and this should be maintained if possible.

The Thermometer

Most tractor engines are fitted with a *thermometer,* the dial of which is conveniently placed for the operator to see, and it gives an indication of the temperature of the water in the cooling system.

Anti-freeze

A water-cooled engine which has to stand or operate in temperatures of freezing point and below must have a proprietry anti-freeze solution mixed with the water. This will prevent damage to the radiator and engine. When water freezes it expands with a considerable force, fully capable of fracturing cast-iron cylinder blocks, cylinder heads or radiators.

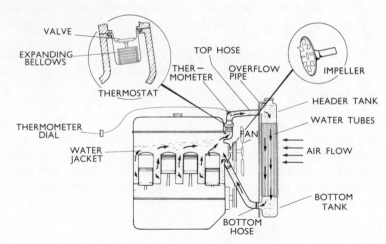

Fig. 16. A water-cooling system.

(Draining a modern cooling system is *not* recommended as protection against frost damage, as a small pocket of water which has not drained properly can easily crack a casting.)

The anti-freeze chemicals lower the freezing point of the water that is treated therefore preventing it from freezing. However, it is necessary to use the correct amount of anti-freeze so that adequate protection is given to the cooling system. The amount to use depends on the number of degrees of frost one can expect during any winter, but usually a 25% solution gives sufficient protection in this country. A gallon of a 25% solution will consist of 6 pints of water and 2 pints of anti-freeze chemical. This will give protection against approximately 30° of frost.

The chemicals in anti-freeze have the effect of giving the water greater penetrating qualities. This means that leaks may develop in the cooling system which were not previously there. It is usually necessary to tighten up all hose connections after the solution has been put in, also the cylinder head bolts must be re-torqued to correct pressure to prevent leaks past the cylinder head gasket.

CHAPTER 3

The Lubrication System

DURING the manufacture of engine parts, a lot of care is taken to make sure that metal surfaces which have to slide or rotate in contact with each other are finished as smoothly as possible. However, the surfaces are never completely smooth, even if they may look or feel smooth. The surface will show roughness when examined under a microscope.

If two such dry surfaces are rubbed together they will get hot, and this heat is caused by *friction*. In fact if they have enough pressure applied and they are rubbed fast enough together they will eventually become hot enough to melt on the surfaces and weld together. The production of such heat can easily be demonstrated by the reader if he rubs his hands smartly back and forth across a table surface.

The effects of friction can be serious and it must be reduced as much as possible where surfaces normally are moving in relation to each other. We do not always want to reduce friction, however; on some mechanisms, such as a clutch or brake, we try to increase it and to do so we use materials that will grip.

The main object of using a lubricating oil is to put a film of oil between the working surfaces of the metal parts. By doing this, friction and wear are reduced because the film of oil keeps the surfaces apart. Some heat is still produced because friction is not entirely eliminated and the oil also helps to cool the metals.

Classification of Oils

Lubricating oils are classified according to their *viscosity* (thickness) and are given a number which tells us the grade of oil, and in some cases the type. This classification of oils was adopted by the Society of Automotive Engineers and it is useful in that it gives the user a means of selecting the grade of oil suitable for his engine. A suitable grade of oil is always recommended by the makers of the engine.

The viscosity of an oil to be used in a particular engine depends on many things and it does not follow that an oil used in engine A can also be used in engine B. It is always wise to follow a manufacturer's recommendation because when he decides to recommend a particular oil for use in his engine, he has considered all features of its design and operation.

Oils of the following S.A.E. numbers are generally *engine oils* S.A.E. 10; 20; 30; 40: whilst S.A.E. 50; 70; 90; 140 are *gear and transmission oils.* The lower numbers indicate low viscosity oil more suited to cold climates, or engines which may stop and start continually, therefore never reaching proper working temperature.

The higher numbers indicate high viscosity more suited to hot climates, or engines continually running in hot conditions.

Detergent Oils

A detergent oil is an oil that contains *additives* which prevent the formation of carbon and lacquer inside an engine. This type of oil must be used for the lubrication of a diesel engine because when diesel fuel burns in the combustion chamber it has a tendency to produce sediments which stick to the working parts of the engine. These sediments collect on pistons, piston rings and cylinder walls where they harden and cause scratching between the pistons and cylinder walls, resulting in increased wear. The additives prevent this from happening by holding in suspension the sticky particles which would otherwise settle on the pistons, etc.

Soot and carbon, which is also produced when fuel and lubricating oil burn within the engine, is also held in the oil and this has the effect of turning the oil black in colour. This is no indication that the oil requires changing.

What is sometimes referred to as an H.D. oil or heavy duty oil is an oil that is *fully* detergent. In other words, it contains additives that not only prevent the formation of sticky deposits but also prevent oxidation and corrosion in the engine.

Detergent oils can be used in petrol engines and will have the same effect as in a diesel engine in preventing sludge and carbon formation. However, if one intends to use it in any engine that has previously run on a non-detergent oil, it is important that a correct change-over procedure is carried out.

Multi-grade Oils

These oils have special additives which reverse the normal tendency of an oil to thicken when cold and thin out when hot. They are numbered S.A.E. 10-30 or 20-50 and are called Viscostatic or Universal engine oils, more suited to petrol or propane engines.

Universal Oils

These oils are produced to reduce the number of different types of oils needed to serve all types of tractors and implements, covering engines, transmissions, hydraulic systems, power steering, etc.

They also eliminate operator misunderstanding by using the wrong type of oil for any specific job.

These oils are of a multi-grade nature with suitable additives to combat engine contaminants, gearbox stresses, anti-frothing for hydraulic systems and power steering.

Changing to Detergent Oils

The use of non-detergent oils causes the formation of carbon and sludge inside an engine. The carbon forms in the combustion chamber, on pistons, valves, and cylinders. Sludge forms and sticks to the walls of the crankcase and in oil galleries.

Because a detergent oil has a cleaning action it will eventually loosen the carbon and sludge deposit and take it in suspension in the oil. This will not happen immediately the change over is made but will take place over quite a few hours running of the engine. Large particles of carbon will be loosened and free to circulate with the oil so that until these particles are completely broken down and taken in suspension there will be a risk that oil galleries become blocked. This could result in oil starvation of bearings which would be ruined in a very short time.

For this reason if a change over from non-detergent to detergent oil has to be carried out, certain precautions must be taken. These entail flushing the engine, running it for short periods between oil changes, and changes of oil filters. Recommended change-over procedures can usually be obtained from the manufacturers of the oil.

Lubricating Systems

Three methods of lubricating engines are commonly used: *splash lubrication, force-feed lubrication* and *petroil lubrication.*

The first is more generally used in small single-cylinder air-cooled engines, although at one time it was used in tractor engines, when they ran at slower speeds than is common now. Force feed lubrication is preferred in all modern diesel and petrol engines. The last method, petroil lubrication, is common to small two-stroke engines, and consists of mixing a small quantity of oil with the petrol. The ratios of oil to petrol for two-stroke engines vary from 16 parts of petrol to 1 part of oil for the larger two-stroke engine, and 24 parts of petrol to 1 part of oil for the smaller two-stroke engines.

Splash Feed Lubrication

A splash system of lubrication is quite simple and, provided that the oil levels are not neglected, nor the periodic oil changes, it is also quite reliable.

As the name of the system suggests, lubrication is achieved by oil splash and in order that this splashing may take place a "dipper" is provided on the end of the big-end cap. When the engine is running, this dipper dips into a narrow trough of oil which is positioned directly beneath it, and throws or splashes oil up on to the working parts. The big end, little end, cylinder, crankshaft bearings, camshaft, timing gears, all being inside the engine crankcase, are lubricated by splash or spray.

Figure 17 shows the essential parts of a splash lubrication system.

It is very important that oil levels should be maintained at the correct position in this system so that the oil trough always holds a sufficient depth of oil. As shown in Fig. 17, this trough is kept supplied with oil through a hole in its side.

Force Feed Lubrication

All modern tractor engines are now lubricated by the force feed method which is very reliable but requires a little more attention by the operator than does the splash method.

The essentials of a force feed lubrication system are a pump usually in the form of a *gear pump,* a pump inlet *oil strainer,* a *pressure-relief valve,* an *external filter,* sometimes a *magnetic filter, oilways* to the various bearings and an *oil-pressure gauge.*

Figure 18 shows the layout of these essentials. The path of the oil is as follows:

When the engine is running, the gear pump, which is driven by the camshaft, draws in oil through the inlet strainer and pumps it to the external oil filter. The pump inlet strainer prevents large particles of dirt or metal entering the pump and causing damage, whilst the external filter should thoroughly clean the oil of any small particles of grit or carbon.

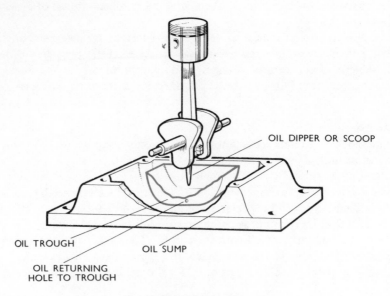

OIL DIPPER OR SCOOP

OIL TROUGH

OIL SUMP

OIL RETURNING
HOLE TO TROUGH

Fig. 17. Splash lubrication.

After passing through the filter, the oil travels through oil galleries and drillings to lubricate the crankshaft bearings, camshaft bearings, big ends, timing gears, valve mechanisms and then returns to the sump for recirculation.

Because the pump speed varies according to the engine speed, the pressure of the oil will also tend to vary and it may in some cases be excessive. When it is excessive, the pressure-relief valve will open and allow the excess oil to return to the sump, whilst oil is still being pumped through the external filter.

A pipe taken from one of the main oil galleries extends to the dashboard of the tractor and is fitted with an oil-pressure gauge. This gauge is a very useful instrument and will give warning of troubles within the engine. The tractor operator must know the correct oil pressure at which his engine should run and he must never try to run it if no pressure is shown but take steps to find out the cause. Oil pressure will vary a little according to engine temperature and condition, but the cause or causes of very high or very low pressures should be investigated immediately. Most modern tractors now fit an oil warning light in place of a pressure gauge. This light glows when the ignition is switched on and is set to go off when the oil pressure builds up to approximately 15 lb per in^2 (1 bar). It is operated by a pressure-sensing switch in the main oil gallery of the engine.

Never operate the engine with the light showing, or engine damage may be the result.

Oil Contamination

Lubricating oil in an engine is contaminated in many ways and it is because of this that it is necessary to change the oil after a certain number of working hours. Moisture, dust, grit, fuel, soot and carbon, and excessive heat all add to the contamination of the oil.

How then does all this foreign matter get into the engine?

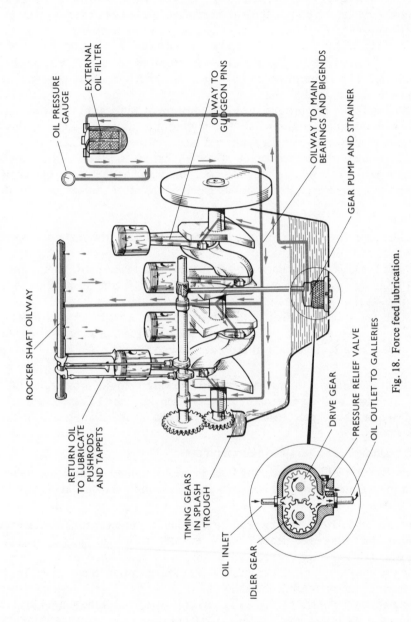

OIL PRESSURE GAUGE

EXTERNAL OIL FILTER

OILWAY TO GUDGEON PINS

OILWAY TO MAIN BEARINGS AND BIGENDS

GEAR PUMP AND STRAINER

ROCKER SHAFT OILWAY

RETURN OIL TO LUBRICATE PUSHRODS AND TAPPETS

TIMING GEARS IN SPLASH TROUGH

OIL INLET

IDLER GEAR

DRIVE GEAR

PRESSURE RELIEF VALVE

OIL OUTLET TO GALLERIES

Fig. 18. Force feed lubrication.

Moisture may be produced within an engine due to condensation and also because moisture is one of the waste products of the burning of fuel.

Dust and grit may enter through the air intake of the engine, particularly if maintenance to the air cleaner has been neglected. Tractors work in all sorts of conditions on farms and dust can be one of their greatest enemies. For this reason they are fitted with efficient *oil bath air cleaners* which should never be neglected. The dust and grit may, of course, be put in by the operator using dirty oil cans and fillers, or leaving caps off oil drums, thus allowing dirt and moisture to enter. Engine oil is also contaminated with unburnt or partial burnt fuel which finds its way past the rings during cold running conditions.

Carbon and soot are produced by the burning of lubricating oil which comes in contact with hot engine parts within the combustion chamber. Some of this carbon and soot is washed down into the sump and causes further contamination.

Metal particles are another source of contamination and these come from wear of the engine parts. This wear is bound to occur but it will happen more quickly if the oil is allowed to become excessively contaminated by the other sources mentioned above.

Oil Filters

Oil filters are fitted within the system to prevent most of the contaminants reaching the working parts of the engine.

The oil filler hole of the engine is often fitted with a coarse mesh filter to prevent the entry of any heavy dirt which may have got into the oil container. This then would be the first stage at which an attempt is made to prevent contamination.

A strainer is fitted around the oil-pump intake to prevent large particles entering the pump. A magnetic filter, which usually takes the form of a magnetized stud, may be fitted at or near to the pump. It protrudes up into the sump and iron or steel particles adhere to it.

External oil filter. All modern engines which employ a pressurized lubrication system are fitted with an external replaceable element filter. These filters are termed **Full Flow** as all the oil must pass through the filter from the pump before it reaches the working parts.

Two main types of filters are in common use.

Replaceable-element type: where the filter element is housed inside a removable casing. This casing is sealed at top and bottom with rubber rings, and held in place by a central bolt.

Replaceable-cartridge type: where the whole unit is unscrewed from the engine and discarded, the fitting area cleaned and a new unit screwed in place by hand pressure sealed with a rubber ring.

Both the above types of filters must be changed each time the engine oil is changed at the manufacturers' recommended hours.

As a guide to the operator, for every gallon (5 litres) of oil the sump holds it should be changed at 100 hours, e.g. 2 gallons = 200 hours.

Oil Changing

Because the engine oil can become contaminated in so many ways, it must be drained, the filter changed and new oil put in. This new oil, as well as lubricating the engine,

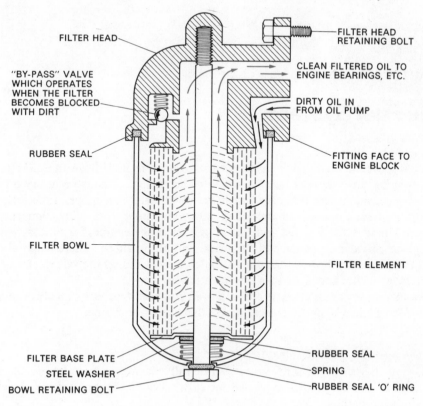

FILTER HEAD

FILTER HEAD
RETAINING BOLT

"BY-PASS" VALVE
WHICH OPERATES
WHEN THE FILTER
BECOMES BLOCKED
WITH DIRT

CLEAN FILTERED OIL TO
ENGINE BEARINGS, ETC.

DIRTY OIL IN
FROM OIL PUMP

RUBBER SEAL

FITTING FACE TO
ENGINE BLOCK

FILTER BOWL

FILTER ELEMENT

FILTER BASE PLATE

RUBBER SEAL

STEEL WASHER

SPRING

BOWL RETAINING BOLT

RUBBER SEAL 'O' RING

Fig. 19. Full flow engine oil filter.

will protect it from damage by contaminants. The oil change must be done when the engine is warm to ensure that the sump is completely drained. Engines that are continually being stopped and started and run for short periods, or engines that run very slowly for long periods, must have the oil changed more frequently, due to excess water and fuel contaminants. Tractor operators should refer to the manufacturer's instruction book as a guide to the oil-change period.

Crankcase Breather

All engines are fitted with a crankcase breather which is necessary to provide an outlet for any build-up of pressure in the crankcase. This pressure may build up due to combustion gases getting past pistons and into the crankcase, and because these gases may contain water vapour they must be allowed to get out of the engine.

The position of the breather and the type will vary according to the engine. Usually it is positioned on top of the rocker-cover on overhead-valve engines, but it may take the form of a pipe coming from the top half of the crankcase. Sometimes the oil filler hole is arranged to form a breather by fitting a filter mesh cover on it. To have a filter on the breather outlet is a common and sensible practice because it helps to prevent dust entering the engine. These filters should be cleaned periodically and wetted with engine oil.

CHAPTER 4

The Fuel System

Air Cleaners

Any fuel system on an engine used for work on farms must have an efficient means of preventing dirt and dust being taken into the engine. This is necessary especially when one considers how filthy the air can get when certain farm operations take place. Examples of such operations are baling, straw chopping, rotavating, lime spreading and combining. Should dust and dirt or grit enter the engine, very rapid wear may take place on valves, pistons and cylinders, and all bearings. There could also be a rapid build up of carbon in the combustion chambers and on the valves. This carbon would glow red hot and upset the running of the engine.

Two main types of air cleaners are employed and both types are capable of removing up to 95% of harmful particles from the air that enters the engine.

The Oil-bath Type, Three-stage Air Cleaner

This type is very popular because it can be fully cleaned and serviced regularly and very cheaply as only new oil is needed for each service. Figure 20 shows a sectioned view of the oil-bath type.

Air enters the pre-cleaner dome tangentially and this sets the air and dirt swirling, thus throwing heavy dirt particles out through the dust-ejection slots. This dome is the first stage. The second stage is when the air passes rapidly down the stack pipe and dust in the air is driven into the oil and trapped. In the third stage the air then passes through an oil-damp wire gauze, trapping any dirt that may have escaped the oil bath. The air then passes out through a rubber hose into the inlet manifold and engine.

Servicing requirements of this type of cleaner are as follows. Clean out first-stage dome regularly, remove oil-bath bowl and clean out stack pipe. Drain the old oil, clean out the bowl and refill to the level with new oil, wash out gauzes in paraffin and blow excess paraffin away with an air line if possible, or allow to drain thoroughly. Re-assemble the filter.

The servicing period can vary according to conditions in which the tractor is working. Finally, check the condition of the hose and the connections between the filter and engine to ensure the hose is not split or loose.

Dry Element-type Air Filters

These filters employ a replaceable element made of wax-paper material. This must be changed when it becomes partly clogged with dirt. It is difficult to give a stated change period due to the type of work and conditions the tractor is operating in. Some tractors have a visual check monitor fitted to the engine induction system, this warns the driver that the filter is becoming clogged.

Some manufacturers recommend cleaning the filter element by various methods to prolong its life. One method suggested is a gentle wash in a detergent and blow out with air line. Another method is to just blow out with the air line from the inside, blowing the dirt out the same way as it enters.

A method of checking the filter element is to lower a bulb through the element to see if the light shows through the element, if so it is fairly clean.

Before attempting any of these methods refer to manufacturer's instructions.

A good guide to a partially blocked air cleaner is to observe the engine's exhaust smoke. Should this appear black it is a good indication that the engine is not receiving its full air supply. A badly neglected air cleaner starves the engine of air which causes excess fuel consumption, power loss and rapid engine wear.

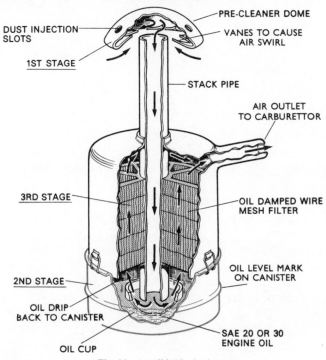

Fig. 20. An oil bath air cleaner.

Petrol Carburettor

During the induction stroke, fuel/air mixture is taken into the cylinder. To make this possible, a device known as a carburettor is used, and it is made so that it will supply the mixture of fuel and air in the correct proportions. These proportions are about 1 part fuel to 15 parts air by weight.

Figure 21 shows the layout of the parts in a simple carburettor. The principle on which it works is as follows:

Fuel is fed from the tank through a pipeline to the carburettor *float chamber* which has the job of maintaining a constant level of fuel. The float chamber is simply a form of container in which there is a small brass float and a *needle valve*. As the fuel feeds into the chamber the float rises and in doing so closes the needle valve which cuts off the fuel supply. As the fuel is used by the engine the float falls, the needle valve opens and fuel is allowed to enter; therefore a constant level is always maintained.

Fuel is stored in the float chamber at atmospheric pressure and a vent ensures that air can flow in and out freely to prevent pressure build up (see Fig. 21).

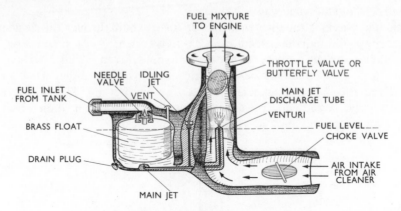

Fig. 21. The layout of the main parts of a simple carburettor.

From the bottom of the float chamber fuel is fed through a jet to a *jet discharge tube* which is positioned in a restricted part of the induction pipe. This *induction pipe* is connected with the cylinders. The actual part which forms the restriction is known as the *venturi*.

The jet discharge tube is positioned so that its end is slightly above the level of the fuel in the float chamber. This prevents fuel from leaking through the outlet when the engine is not running.

When the engine is running, the downward movement of the pistons on the induction strokes causes a depression in the cylinders with the result that air rushes in and in doing so it must pass through the induction pipe in which the jet discharge tube is situated. The movement of the air through the venturi draws fuel from the discharge tube which mixes with the air. The mixture, which is in a vaporized form, enters the engine cylinders through the inlet valves.

This then is basically the principle on which the carburettor works; of course, carburettors are not usually quite as simple as this. A more complicated arrangement is needed because as engine speeds increase and decrease so do the proportions of fuel and air being taken into the engine. In order to provide a correctly balanced mixture at different speeds, devices such as air bleeds, compensating jets, pilot jets and idling jets are used.

Figure 21 shows the feed to an idling jet which supplies the fuel when the engine is ticking over.

Two other parts shown in the illustration are the *choke valve* and the *throttle valve*. These are flat circular discs pivoted inside the induction pipe. The choke valve is on the air cleaner side of the carburettor and the throttle valve is on the engine side. It is often necessary to provide the engine with a rich mixture for starting and the choke valve gives us a means of doing this. When it is closed, or partly closed, the air intake is restricted so that the mixture contains a higher proportion of fuel. The illustration shows the valve fully open.

To control the supply of fuel/air mixture into the engine, the throttle valve is used and on a car this would be interconnected to the accelerator, but on a tractor it is connected to a governor and hand control. To reduce the engine speed it is only necessary to close or partly close this valve.

The Governor

All tractor engines are fitted with a governor which has the job of helping to maintain a constant engine speed once that speed has been set by the hand control.

Figure 22 shows a diagrammatic layout of a mechanical governor.

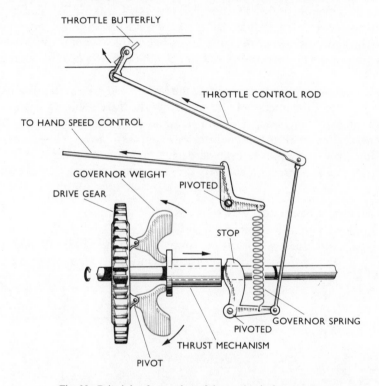

Fig. 22. Principle of operation of the mechanical governor.

When the drive gear rotates, the governor weights are caused to move in the direction shown by the arrows. This movement is caused by centrifugal force and the extent to which the weights move depends on how fast the gear rotates. If the gear rotates at high speed, the weights move further out. As the speed decreases the weights fall. The gear is usually driven by the engine timing gears. The weights are in contact with a thrust mechanism which can slide along the shaft; this mechanism in turn is in contact with a spring-loaded stop device which is connected to the carburettor throttle and the hand speed control.

Now let us consider its action when working.

When the hand speed control is pulled in the direction of the arrow, the end of the small crank to which the governor spring is attached moves up. This causes another two things to happen; the pull on the governor spring moves the throttle control rod upwards and this opens the throttle, giving a faster engine speed, and the faster engine speed causes the weights to move farther out so that they thrust harder on the thrust mechanism. This also causes an increased resistance on the governor spring. The result is that the faster engine speed is maintained.

Now let us consider what happens when, say, a tractor is ploughing.

The speed of the engine is set by the tractor driver with the hand speed control. If when he is ploughing he comes to a patch in the field which is heavy land, the tendency will be for the engine speed to slow down because of the increased load. This decrease in engine speed results in a decrease in the speed of the governor drive gear which in turn causes the governor weights to retract closer to the shaft. This causes the thrust mechanism to move to the left under the action of the stop and governor spring.

Because the hand speed control is in a set position, the governor spring pulls up and in doing so moves the throttle control rod to open the butterfly and maintain the desired speed.

When the load on the engine is reduced, the reverse will take place. For example, at the headland when the plough is lifted out of work, there will be a tendency for the speed of the engine to increase. This will cause the governor weights to fly farther out and push harder on the thrust mechanism. This in turn moves the stop which causes the governor spring to be pulled downwards and in doing so the throttle rod is moved down and this tends to close the butterfly thus again maintaining the set engine speed.

General Layout of a Petrol Fuel System

Figure 23 shows a typical layout of the petrol system as employed on small stationary engines or petrol tractors still in use. It should be noted that ample precautions are taken to ensure that dirt does not enter the system. Small amounts of dust and grit,

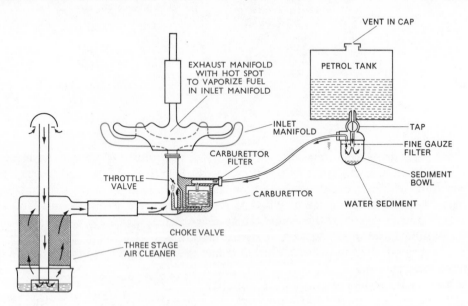

Fig. 23. Layout of the fuel system on a vaporizing oil engine.

etc., in the fuel supply can cause endless trouble until they are removed. The jet outlets in carburettors are quite small and easily choked, so preventing the fuel from discharging properly.

To help prevent such troubles, as many as three fine mesh filters may be used in the fuel line, one in the tank, one in the sediment bowl and another in the carburettor float chamber inlet. The sediment bowl also accumulates any moisture that gets into the fuel. This may get in when condensation takes place within the fuel tank. The bowl should therefore be inspected periodically and cleaned out.

The Diesel Engine Fuel System

The main difference between a petrol engine and a diesel engine is in the method in which the fuel is supplied to the cylinders and ignited.

The diesel engine has no carburettor and no electrical equipment to provide a spark to ignite fuel. Instead, it has a fuel injection pump and fuel injectors. The basic operating cycle is the same as that of the petrol engine except that when the induction stroke takes place, air only is taken into the cylinder through the inlet valve. This air is compressed to a very high degree during the compression stroke and the high compression causes the air to become very hot. At the end of the compression stroke, fuel in the form of a fine mist is sprayed into the cylinder by an injector. The fuel immediately ignites when it comes in contact with the hot air and the power stroke takes place.

The main parts in the diesel engine fuel system are: an air cleaner, a fuel lift pump, fuel filters, fuel injection pump and fuel injectors.

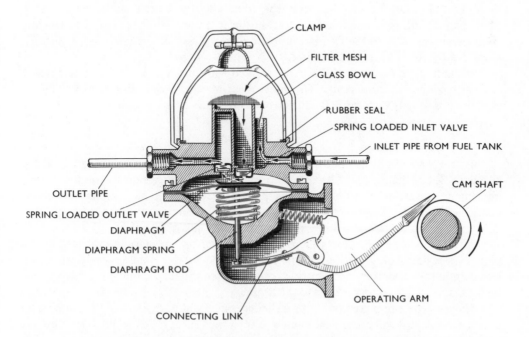

Fig. 24. Section through a fuel lift pump. The arrows show the path of the fuel.

The Air Cleaner

This is the same type of air cleaner as fitted to petrol engines, and it should thoroughly filter all the air taken into the engine. See Fig. 20.

The Fuel Lift Pump

A fuel lift pump is usually fitted to the fuel line of the diesel fuel system. Whether or not a lift pump is fitted usually depends on the position of the fuel tank. Such a pump is necessary on vehicles where the tank is not positioned high enough to maintain a good gravity flow. This is always the case with cars but not always with tractors because the fuel tank is placed high up beside the engine. All petrol engines on tractors usually rely on gravity flow of fuel to the carburettor, but in the case of the diesel fuel system the use of a lift pump provides fuel at a constant pressure to the injection pump. This pressure also helps to push the fuel through filter elements before it gets to the injection pump.

Figure 24 shows a diagram of a lift pump.

It is operated by a cam on the engine camshaft so that it pumps continually as long as the engine is running. A filter gauze is fitted within the pump and it serves as one point where dirt and moisture can be trapped.

Fuel Filters

One or more fuel filters are used to filter the fuel as it is pumped from the lift pump to the injection pump. These filters are most important and their purpose is to ensure that absolutely no dirt of any kind finds its way to the injection pump or injectors. They take the form of a metal container in which there is either a cloth, felt or paper filter element and they are arranged so that the fuel must pass through the element. In this way, any dirt in the fuel is trapped on the surface of the element. Figure 25 shows a diagram of a typical fuel filter.

Eventually these elements will become choked with dirt so it is necessary to replace them at recommended intervals. The interval varies according to the type of elements used but it is usually between 200 and 400 working hours of the engine.

The Fuel Injection Pump

This is a very important and expensive part of the diesel engine. It is expensive because a lot of highly skilled work goes into making a precision instrument. It is important because when in use it must deliver accurately to the engine very small quantities of fuel. Furthermore, in a multi-cylinder engine, the quantities of fuel delivered to each cylinder must be equal and delivered at high pressure.

Two types of fuel injection pumps are in general use on agricultural tractors, a plunger type pump or a distributor type pump. In both cases engine speed is increased by operating the necessary acceleration control which causes the pump mechanisms to feed an increasing quantity of fuel to the engine.

A plunger type pump used on a four-cylinder engine will be made up of four pumping units and each unit will pump fuel to an injector which sprays the fuel into the cylinder, whilst the distributor pump makes use of a single pumping and distributing rotor which feeds each injector in turn.

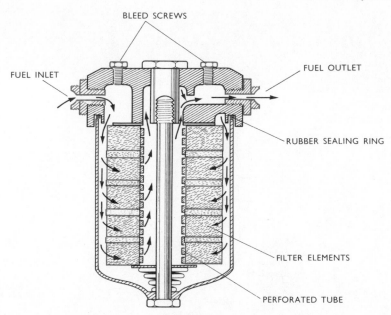

Fig. 25. Fuel filter for a diesel fuel system. The arrows show the path of the fuel.

The Fuel Injector

The fuel injector is another very accurately made part of the diesel fuel system and its purpose is to inject into the cylinder the small quantities of fuel.

This fuel must be broken up into a very fine spray and this is achieved partly because it is forced through minute holes in the end of the injector and partly because the pressure which forces it through these holes is very high. The holes may be no more than 0·2 mm (8/1000 in.) in diameter and the pressure in the region of 140 kg/cm² (2000 lb/in² approx.).

General Layout of the Fuel System

Figure 26 shows the general layout of a diesel engine fuel system, and the path of the fuel is indicated by the arrows. It will be seen that, as with the petrol engine, ample precautions are taken to ensure that only clean fuel reaches the injection pump and injectors. This is of greater importance in a diesel engine than in the other types of engine, because the smallest particles of dirt can cause much wear to the highly machined parts of the pump and injectors.

Let us trace the fuel to the fuel injector. Firstly, the air supplied to the engine must be clean and this is catered for by the oil bath type air cleaner.

The fuel in the tank should always be filtered in whenever refuelling is carried out. From the tank the fuel passes through the tap into a sediment or filter bowl. Such a bowl is not always fitted, but one is shown in this case as being an additional means of cleaning the fuel.

From the filter bowl the fuel is fed by gravity down to the lift pump in which there is another fine mesh filter. The lift pump pumps the fuel up and through the main fuel

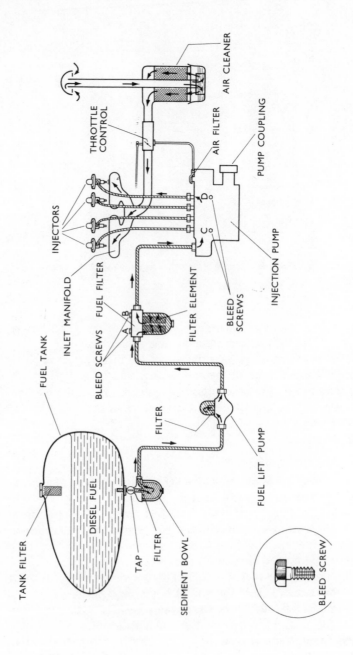

Fig. 26. Layout of a fuel system on a diesel engine.

filter. On some tractors there may be two of these filters. From this point the fuel passes to the injection pump which meters it and pumps it through the injector which sprays the fuel into the cylinder in which compression of the air has taken place. In the illustration it is shown as fuel being injected to the first cylinder.

The injection pump is coupled to the engine's timing gears by the pump coupling and it is timed so that fuel is injected into the correct cylinder at the correct time.

When it is necessary to increase the speed of the diesel engine, the operator opens the throttle control which moves a mechanism within the injection pump and causes more fuel to be fed to each cylinder. Figure 26 shows a pipe connection between the injection pump and the throttle control. This is an air connection and is only present if a pneumatic governor is fitted. To stop the engine a fuel cut-out control is operated and this moves the mechanism which in this case prevents fuel from being delivered to the injectors.

The Governor

It is essential that the speed of a diesel engine can be controlled by some form of governor. This will prevent engine damage by over-revving and also help maintain close speed control for any particular machine the engine may be driving. There are three main types of governor in use:

Mechanical governor. These are the most popular as they give very quick response to any load changes therefore maintaining reasonable constant revs of the engine. Both in-line and rotary diesel injector pumps can be fitted with mechanical governor control directly coupled to the Helix fuel control of the injector pump. The principle of operation of a mechanical governor on a petrol engine is shown in Fig. 21.

Pneumatic governor. These have been fitted in inline injector pumps but tend to respond more slowly to load change therefore causing a considerable drop or gain in engine revs before correcting and settling. This causes speed variation on a machine operating on power drive.

Hydraulic governor. This term is misleading as it implies the use of hydraulic pumps and rams to the layman, but in fact it is quite a simple system built into D.P.A. rotary injection pumps. The revs of the engine affects transfer pump output pressure of fuel which acts on a metering valve to control engine revs according to hand throttle setting. This type of governor is slow on response and is found on fork-lift truck engines, where close speed control is not essential.

Bleeding the Fuel System

If air gets into the fuel system it means that as air pockets reach the injector, no fuel will be injected into the cylinder. Either the engine will run erratically or it will stop entirely, depending on how much air there is in the system. How can air get into the fuel system? It will get in if the driver allows the fuel tank and all the fuel lines to run dry when working. It will also get in when any servicing is done to the fuel system such as cleaning out the fuel lift pump or the replacement of the main filter element. In fact, air will get into the system whenever the fuel line is dismantled for any reason.

To remove the air in the system it is necessary to carry out the operation known as "bleeding".

The method of doing this varies slightly with makes of tractor and the correct method can be got from the instruction book, but the principle can be explained with reference to Fig. 26. Two bleed screws are shown fitted to the top of the main filter casing and another two on the injection pump. Let us assume that the tractor driver runs his engine until the fuel tank is dry and the engine eventually stops. The system would be primed as follows.

The fuel tank is first filled, and with the fuel tap open, fuel will flow by gravity to the lift pump. It may be necessary to slacken the clamp holding the sediment bowl in order to allow the fuel to flow in freely. The next step is to slacken bleed screw A, and operating the hand priming lever on the lift pump, pump fuel through to the filter until fuel without any air bubbles is forced out of the screw. Then tighten up the bleed screw. The same procedure is carried out with bleed screw B. This then has ensured that all air has been removed up to this point.

Now slacken bleed screw C and continue pumping until all air bubbles are again removed, tighten the screw then repeat the operation at bleed screw D. We should now have all air removed up to and including the injector pump.

To remove the air from the injector feed pipes it is necessary to slacken each pipe in turn at the injector end and turn over the engine with the starter. In this way the injector pump will force out the air. The pipes are retightened when fuel is seen to come out. The engine should now start.

The inset of Fig. 26 shows the usual construction of a bleed screw. It has a hole drilled down its centre and then out to the side. This arrangement makes it unnecessary to remove the screw entirely when bleeding the system. Providing it is unscrewed just sufficiently for the hole in the side to allow fuel to flow out, the bleeding can be carried out. This type of screw is used to reduce the risk of getting dirt into the fuel line.

The Handling of Diesel Fuel

The importance of the clean handling of diesel fuel can never be over-emphasized if running troubles and wear in the engine's injection equipment is to be kept to a minimum.

This equipment, as already pointed out, is very precisely made and even the finest particles of dirt coming in contact with the working parts can cause extensive wear. Efficient filtering helps to prevent this but dirt may still enter if careless servicing or bleeding is carried out.

Even though precautions are taken to filter the fuel on the tractor there is no point in giving these filters more work than necessary, therefore, clean handling of fuel starts where it is stored. No fuel filter can prevent water or liquid acids from passing through it, both these liquids can do extensive damage to the precision parts of the injection pump and injectors. Figure 27 shows a suitable layout for the storing of diesel fuel and the important features of it are as follows.

(a) The tank must not be made of galvanized iron because the action of the diesel fuel tends to remove the zinc coating thus contaminating the fuel.

(b) The tank is set up on pillars at such a height as to allow easy filling of the tractors.

(c) The tank slopes to the rear. This allows sediments to collect at this point and away from the outlet.

(d) A sludge tap is fitted at the rear to drain off the sediment at intervals.

(e) A downturned air entry pipe is fitted to the top of the tank. This prevents the entry of dirt but allows air in whilst fuel is drawn off.

(f) The pipe through which the tank is filled is covered with a screw cap.

(g) A filter is fitted to the tank end of the refuelling pipe.

(h) The valve on the refuelling pipe is kept covered in a box beneath the tank.

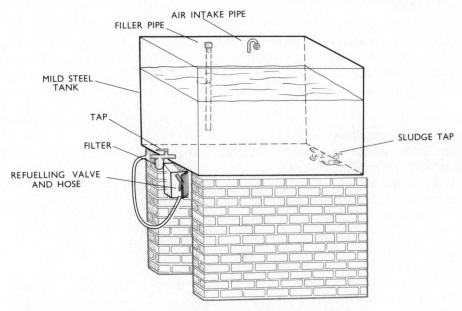

Fig. 27. A suitable arrangement for storing diesel fuel.

The arrangement shown will go a long way to ensuring that only clean fuel goes into the tractor but proper and careful servicing of the tractor's fuel system is also necessary if more dirt is not to get to the vital parts.

CHAPTER 5

The Ignition System

TO PROVIDE an electric spark to ignite the fuel/air mixture in a spark ignition engine, one of two arrangements may be used. These are (1) *magneto ignition* and (2) *battery and coil ignition*. Both are used quite widely on tractor engines and magneto ignition is also used on stationary engines. The coil ignition is perhaps the more elaborate.

Battery and Coil Ignition

Some knowledge of electricity is necessary to understand how this system works. This information is given in Chapter 10.

The first important unit of the coil ignition system is the battery. This supplies the initial current to operate the system, therefore it is important that it is kept in good order by proper maintenance. The maintenance will consist of

1. Ensuring that the battery is kept clean and dry externally.
2. Ensuring that it is kept adequately topped up with distilled water, e.g. the cell plates within the battery must be just covered.
3. The battery terminals must be kept tight and free from corrosion. A smear of vaseline on the terminals will help prevent corrosion.
4. The battery must be rigidly fixed in position on the tractor to prevent movement and possible spillage or damage.
5. The battery must always be kept in a state of charge.

The Coil Ignition Circuit

The electrical spark which ignites the fuel mixture must take place across the points of a spark plug, and this plug protrudes into the combustion chamber where the fuel is compressed. This means that the spark must jump across the points against the high pressure within the cylinder. The 12-V battery on its own is not capable of causing a spark under such conditions; the voltage must be increased to something like 7000-10,000 V. This is achieved by the use of a *high-tension ignition coil*.

Figure 28 shows the layout of a coil ignition system and the method in which the spark is produced at the plug is as follows:

There are two separate electrical circuits in the system: the *primary circuit* and the *secondary circuit*. On the illustration the black arrows show the current flow in the primary circuit, and the red arrows show the flow in the secondary circuit.

When the ignition switch is closed, current flows from the battery into the ignition coil where it passes through the primary winding. After passing through this winding the current then passes to the distributor which houses a contact breaker assembly which is a form of automatic switch. This contact breaker assembly is also in the

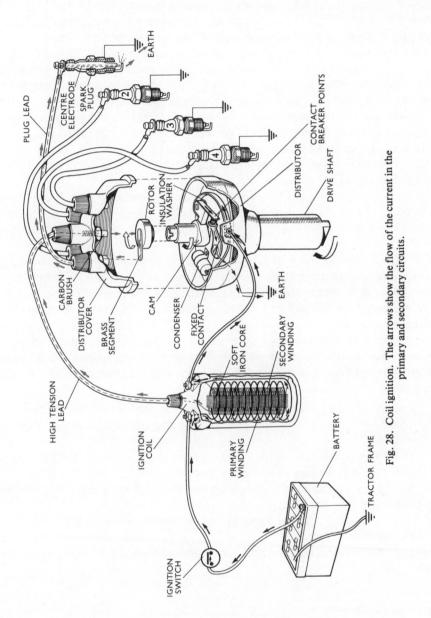

EARTH

PLUG LEAD

CENTRE
ELECTRODE

SPARK
PLUG

CONTACT
BREAKER POINTS

DISTRIBUTOR

DRIVE SHAFT

ROTOR

INSULATION
WASHER

CARBON
BRUSH

DISTRIBUTOR
COVER

BRASS
SEGMENT

CAM

CONDENSER

FIXED
CONTACT

EARTH

HIGH TENSION
LEAD

SOFT
IRON CORE

SECONDARY
WINDING

IGNITION
COIL

PRIMARY
WINDING

BATTERY

TRACTOR FRAME

IGNITION
SWITCH

Fig. 28. Coil ignition. The arrows show the flow of the current in the
primary and secondary circuits.

primary circuit and, as the cam rotates, the contact points will open and close. Now the current coming from the coil passes, as shown by the arrows, through the points and back to earth, thus completing the circuit. But this only happens when the points are closed. The flow of current also causes a magnetic field to be set up around the primary winding. The moving contact is insulated from the distributor or body by an insulating washer, therefore the current must go to earth through the fixed contact. Notice also that a small part known as a *condenser* is in the primary circuit.

Now if this circuit is broken, in other words the current cannot flow to earth in its usual way, the magnetic field surrounding the primary windings collapses and in doing so cuts across the windings in the secondary. This produces a small current at high voltage which flows through the secondary circuit. The voltage may be up to 10,000 V.

The red arrows on the diagram show where this current flows. It leaves the ignition coil through the high tension lead and goes to the centre of the distributor cap where the carbon brush is in contact with the metal segment on the rotor. If this segment is opposite a segment in the distributor cap, the current will pass from one to the other and then through the plug lead.

It passes through the plug lead to the spark plug where it goes down the centre electrode, jumps across the gap creating a spark and then to earth to complete the circuit. The current is able to jump across the gap because of the high voltage pushing it.

When the engine is running, the drive shaft of the distributor is rotated by the engine camshaft and as this drive shaft has the cam and rotor fixed to it, these will also rotate. This means that as the shaft rotates, the cam will operate the contact breaker points, causing them to open and close, and the rotor will pass each segment on the distributor cap in turn.

Now the ignition system of an engine is timed so that when a piston is at the top of its compression stroke, the rotor segment is opposite to the segment of the plug lead leading to that cylinder, and the contact breaker points are also open. So the following takes place:

1. The ignition switch is closed and the engine is turned over by the starter.
2. Current flowing through the primary circuit creates a magnetic field in the windings.
3. The contact breaker points are opened by the cam; this breaks the primary circuit causing the high voltage to be induced into the secondary circuit.
4. The rotor is opposite a segment leading to a plug in the cylinder where compression is taking place.
5. A spark is created across the plug points and the mixture is ignited.
6. The engine starts to run.

This sequence of events occurs when each piston in turn is at the top of the compression stroke.

Purpose of the Condenser

This small part has an important job to do in the primary circuit. When the current flows through the primary circuit, immediately the contact breaker points open to break the circuit, the current tries to carry on by jumping across these points. If this was allowed to happen the points would soon become badly burnt. The condenser prevents much of this from happening because it is so arranged in the circuit that when the points open, instead of the current jumping across them it takes an easier path into the condenser where it is temporarily stored.

The unit of electrical capacity by which condensers are measured is the microfarad. Condensers for coil ignition are 0·2 MF, and for magnetos 0·01 MF.

Magneto Ignition

The magneto is a self-contained unit driven by the engine and it generates electricity which is transformed to a high voltage when required.

The usual arrangement on tractors is to drive the magneto by the timing gears of the engine. Because it is capable of generating its own electricity, the use of a battery is not required unless the tractor is equipped with lighting.

In the magneto the primary and secondary coils are wound on to an armature shaft which rotates between the poles of a horseshoe magnet. In this way current is produced to flow through the circuits. An alternative arrangement is to have stationary coils and a rotating magnet.

The operation of the magneto is almost identical to that of the coil ignition system, except that the initial primary current is not produced by the battery but is induced by the relative movement of coil and magnet.

The essential parts of a typical magneto ignition system are as follows:

1. Permanent magnet
2. Primary winding
3. Secondary winding
4. Contact breaker
5. Condenser
6. Distributor
7. Magneto body
8. Plug leads
9. Spark plugs

Figure 29 shows a semi-diagrammatic layout of a rotating magnet and stationary coil type magneto ignition system. Its operation is as follows:

When the engine drive gear is rotated, and this would happen when the starting handle was cranked and also, of course, when the engine is running, the rotating magnet rotates between a pair of pole shoes. These pole shoes are bridged by a soft iron core around which the coils are wound. Rotation of the magnet causes a magnetic flux to pass through the pole shoes and also through the iron core. When this happens, the lines of force in the magnetic field cut through the windings in the primary coil causing a current to flow in the primary circuit. The illustration shows the circuit and the black arrows indicate the path it takes. The circuit is complete when the contact breaker points are closed. The cam opens the contact breaker points as its rotates. Opening of the points breaks the circuit and this induces a high voltage in the secondary windings and the current follows along the path shown by the red arrows on the illustration. That is, to the secondary pick-up, through the rotating contact which is timed to be opposite a segment in contact with a plug lead, then through the plug lead to the centre electrode of the spark plug. It then jumps across to the point on the plug

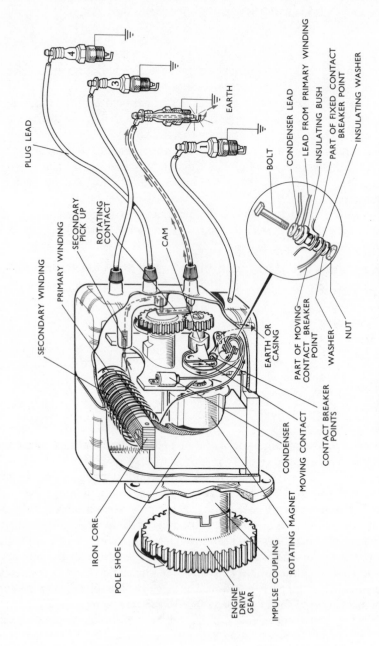

Fig. 29. Magneto ignition. The arrows show the flow of the current in the primary and secondary circuits.

body and goes to earth. A spark is thus created which ignites the fuel mixture in the engine cylinder.

In this ignition system, as in the coil ignition system, the condenser serves the purpose of absorbing the current which tends to jump across the contact breaker points.

The Impulse Coupling

An increase of speed of rotation of the magnet near to the coils will also cause an increase in voltage. Now when an engine is turned over by hand, the magneto is not driven at sufficient speed to provide a suitable voltage to cause the spark across the plug points. To overcome this, a device known as an impulse starter is used. This is fitted to the shaft of the rotating magnet and is coupled to the engine drive gear. It is so made that when the engine drive gear rotates the magneto shaft remains stationary whilst a spring within the impulse starter winds up. At a predetermined point the spring is released and the magneto shaft turns quickly enough to give the necessary spark for starting. The mechanisms of the impulse starter are held out of engagement by centrifugal force once the engine starts.

The Charging Circuit

The electrical system of a tractor is made up of a number of electrical circuits. For instance, on the spark ignition tractor not using a magneto, there will be the coil ignition circuit, the starter circuit, and if lights are fitted there will be a lighting circuit. The diesel tractor has a starting circuit and may also have a lighting circuit.

All these electrical circuits must get their supply of electricity from the battery and because the battery cannot go on supplying this indefinitely without becoming exhausted, we must have some means of restoring it. To do this we use a device known as a dynamo.

The Dynamo

The dynamo is driven by the engine and it supplies direct current (d.c.) to the battery. An explanation of how the dynamo works is given in Chapter 10, therefore at this stage it is only necessary to mention some aspects of maintenance.

The dynamo is mounted on the side of the engine and in such a position that it can be driven by the belt which also drives the radiator fan. It is also conveniently arranged so that the dynamo can be moved to set the tension of the belt when necessary. This belt tension should always be kept correct because if it runs too slack slipping will occur and this will result in a reduced speed of the dynamo pulley and also the radiator fan. The ultimate result of this could be that the cooling system overheats or the dynamo may not run at sufficient speed to give full charge. If the belt is run too tight it is likely to stretch and also to cause excessive wear to the dynamo bearing.

Some dynamos are fitted with a lubricator which requires a few drops of light oil periodically, but many are now pre-lubricated and require no further attention. The only other maintenance likely to be done to the dynamo is the fitting of new carbon brushes and undercutting of the commutator but this work is better carried out by a skilled mechanic.

Dynamo Regulation

The speed at which the dynamo operates will depend on the speed of the engine. Therefore, because dynamo output depends on the engine speed, as this varies so will the dynamo output. In fact the position can arise where, when the engine is running very slowly, the dynamo is producing less voltage than is actually in the battery. In such a case, and with the battery being connected to the dynamo, it is possible for the battery to discharge itself through the dynamo windings. This, of course, cannot be allowed to take place, therefore a device known as a *cut-out* is fitted in the charging circuit. The cut-out is a form of magnetic switch which opens and closes according to the voltage produced and connects and disconnects the battery as required.

The position can also arise where, when the engine is running at high speed, the voltage produced is too high and may damage the battery and other parts of the electrical system. To overcome this problem a device known as a *voltage regulator* is fitted in the circuit. This is another switch device similar to the cut-out but in this case its function is to cause a reduction in the voltage produced.

The cut-out and voltage regulator are quite complicated devices not to be tampered with. Both are housed in a single container which is usually sealed to prevent access. Figure 30 shows the position of this unit, and also of the ammeter in the charging circuit. This instrument, which would be fitted on the dashboard of the tractor, shows whether or not the charging system is working properly. For example, if, say, the dynamo brushes were excessively worn and were not picking up the current, or if the lights were on whilst the engine was not running the ammeter would show a discharge, as the instrument needle reads zero when it is vertical at the mid-point on the scale.

The Starter Motor

Tractor engines can be started by hand by turning over the engine with a cranking handle, in fact, where magneto ignition is used this is the usual method of starting. Turning an engine over by hand can sometimes require quite an effort, especially when

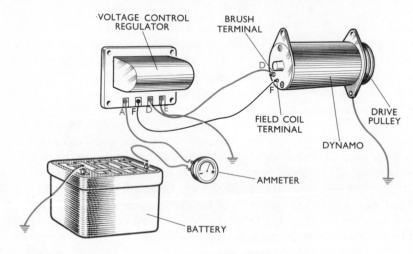

Fig. 30.

attempting to start from cold. Therefore, where a battery is fitted to the tractor, it is also normal to have an electric starter. This gives push-button starting and it is used on spark ignition tractors and diesel tractors.

The principles of operation of a starter motor are given in Chapter 10. At this point we can see how it engages with and turns over the engine.

How the Starter Turns the Engine

The starter is heavily constructed to enable it to carry the heavy current required to give it good turning power, for it must have power enough to move the engine. The starter motor shaft around which the armature coils are built extends out of the yoke or casing. To the end of this shaft is fitted the mechanical device which engages with the engine flywheel to rotate it. Figure 31 shows the arrangement.

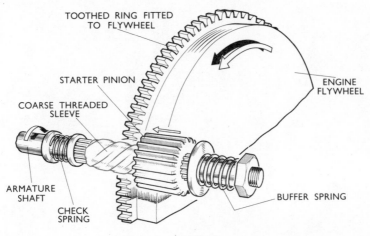

Fig. 31.

When the starter switch is operated, current flows through the armature and field coils and the armature rotates quickly. This also causes the coarse threaded sleeve to turn because it is splined on to the armature shaft. The starter pinion is internally threaded and the effect of rotation of the threaded sleeve is to cause the pinion to move in a straight line into mesh with the flywheel toothed ring. This straight-line movement of the pinion is due to its unwillingness to rotate at speed with the sleeve. It therefore moves straight into mesh with the flywheel ring and also comes up against the check spring. This check prevents further straight-line movement of the pinion, therefore it must now rotate and in doing so it turns the flywheel so that the engine may start. As soon as the engine starts and the starter switch is disengaged, the flywheel toothed ring will be driving the pinion and the pinion is thus screwed back off the sleeve and against the buffer spring.

Many modern diesel tractors now use "pre-engaged" type starters, where the pinion is moved into mesh with the flywheel ring gear before the starter turns. There are two main types of pinion engaging mechanisms:

Manual pre-engaged type, where the pinion is moved into engagement by a lever, which is pushed down by the operator. As the lever reaches the end of its movement it operates a push-button switch which in turn operates the starter solenoid that passes heavy current to the motor to turn the engine.

Electrical pre-engaged type. On pressing a dashboard switch or turning a key, a powerful solenoid mounted directly on top of the starter motor works a lever to engage the starter pinion with the flywheel ring gear. At the end of the movement a switch is operated to pass current to the motor to turn the engine.

Both of the above types are popular as they reduce pinion and flywheel ring gear damage and wear.

CHAPTER 6

The Transmission System

THE transmission system of the tractor is made up of a number of units each of which has a separate and important part to play in transferring the power of the engine to the rear wheels, so that the tractor will move and be capable of doing work.

Firstly, it is necessary to have some device that can be used easily and conveniently to connect the engine to the rest of the transmission system. It must be just as easy to disconnect the engine when required so that it may continue to run without moving the tractor. Furthermore, when the engine is required to drive the transmission, the connecting device must be able to take up the drive gradually and smoothly.

The *clutch* performs this job and it is the first unit in the transmission system.

The Clutch

Most tractors are fitted with a dual clutch which not only provides the means of connecting the engine to the tractor transmission but also provides power from the engine to what is known as a "live" power take-off shaft. Fig. 33 shows a section through a dual clutch. A more detailed explanation of this clutch is dealt with on page 73 under the heading of The Live P.t.o. At this point we are mainly concerned with the operation of a single clutch.

This is known as a "single-plate dry friction clutch" and it is directly attached to the engine flywheel so that it rotates with it. Figure 32 shows the principle on which it works.

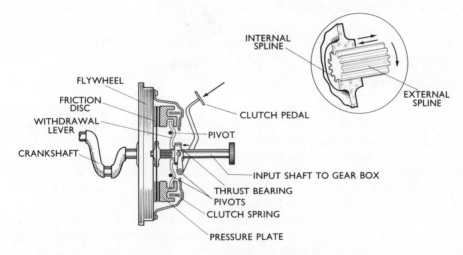

Fig. 32. The single-plate dry friction clutch.

A friction disc, which is faced with a friction type material, is splined on to the clutch shaft. Splines are formed simply by cutting rectangular shaped grooves out along a portion of the shaft. If we have internal splines cut in the centre of the clutch plate, these can be mated on to the shaft so that it will slide along the shaft, yet rotate with it. The inset of Fig. 32 shows this construction.

The end of the input shaft is fitted into a bearing in the centre of the flywheel and the plate is positioned between the face of the flywheel and a spring-loaded pressure plate, Fig. 32. In order to transmit the drive from the engine flywheel to the clutch shaft which goes directly into the gearbox, it is only necessary to allow the clutch plate to be gripped between the faces of the flywheel and pressure plate. The heavy spring pressure is sufficient to do this and form a solid drive to the shaft. On an actual tractor clutch there may be as many as twelve heavy coil springs used to apply the pressure. To disconnect the drive it is only necessary to withdraw the pressure plate against the spring pressure. This is done by pushing down the foot pedal which pushes forward the thrust bearing causing the outer ends of the withdrawal levers to push back the pressure plate against the pressure of the clutch springs, so that the clutch disc is no longer gripped between the pressure plate and the flywheel and the drive to the clutch shaft ceases.

Misuse, by excessive slipping of the clutch plate between the flywheel and pressure plate, will result in severe wear and damage to the friction material on the plate itself. When applying pressure to the foot pedal to release the clutch plate, a point will be reached where slipping can occur, that is, when all the plate surfaces touch each other but are not gripped. To hold the mechanisms in this position for prolonged periods can result in sufficient heat being produced, by friction, to burn the lining on the clutch plate and perhaps also distort the plate. For this reason, slipping of the clutch must be avoided if possible.

Although this type of clutch is perhaps most commonly used, there are, of course, other types. One such type is known as the *multi-plate clutch* and it is frequently used as a steering mechanism on track-laying tractors. In this clutch more than one plate is used, there may be as many as nine or more. The arrangement of them is such that every other one would be splined on to the shaft whilst the others would be fixed to the driving member.

The principle of operation is the same as in a single-plate clutch, but in this case all the plates are sandwiched together by spring pressure.

The Gearbox

The second unit in the transmission system is the gearbox, and one of its main purposes is to reduce the speed of the drive from the engine crankshaft before the drive is applied to the rear wheels of the tractor. It is necessary to reduce the speed of the engine before it reaches the land wheels, otherwise they would be travelling far too fast. The gearbox is not the only unit in the transmission which brings about a speed reduction between the engine and rear wheels, but it is the only one where the speed ratio can be altered easily and conveniently as required.

The power that the engine develops increases as the speed of the engine increases. This being so, if a tractor is working with the engine set at full throttle, it will be developing its full power. If a situation arises where more power is required, and it cannot be given, two things will happen. Firstly, the engine will begin to slow down

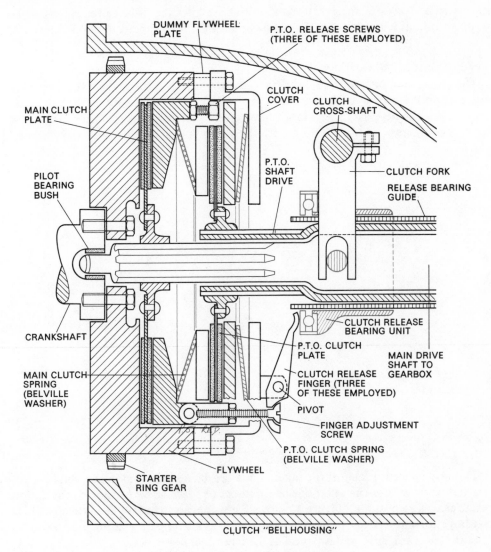

DUMMY FLYWHEEL PLATE

P.T.O. RELEASE SCREWS (THREE OF THESE EMPLOYED)

CLUTCH COVER

CLUTCH CROSS-SHAFT

MAIN CLUTCH PLATE

PILOT BEARING BUSH

P.T.O. SHAFT DRIVE

CLUTCH FORK

RELEASE BEARING GUIDE

CRANKSHAFT

CLUTCH RELEASE BEARING UNIT

P.T.O. CLUTCH PLATE

MAIN DRIVE SHAFT TO GEARBOX

MAIN CLUTCH SPRING (BELVILLE WASHER)

CLUTCH RELEASE FINGER (THREE OF THESE EMPLOYED)

PIVOT

FINGER ADJUSTMENT SCREW

P.T.O. CLUTCH SPRING (BELVILLE WASHER)

FLYWHEEL

STARTER RING GEAR

CLUTCH "BELLHOUSING"

Fig. 33. A section through a Dual Clutch.

because it is being overloaded, and secondly, because of the slowing down of the engine, the power will decrease. This will result in a further loss of speed and power until eventually the engine will stall and stop altogether. When this happens an experienced driver instinctively changes into a lower gear and overcomes the need for extra pull, as the changing down means that if the engine speed remains constant the tractor will now travel forward a lesser distance than it did before during the same number of engine revolutions. The gearbox does not increase power in any way, but it does make the power of the engine more easily applied to the work that the tractor may be doing. This is done by changing gear and allowing the tractor wheels to rotate still slower in relation to the engine speed.

The gearbox serves other purposes, of course; it provides a means of allowing the tractor to be reversed when required and it also allows the drive to the wheels to be stopped without having to stop the engine or keep the foot applied to the clutch pedal.

Gearbox Layout and Operation

The principle of operation of gears is discussed in Chapter 9 and the tractor gearbox operates on these same principles in that suitable pairs of gears are meshed together and provide the necessary speed ratios between the engine and rear wheels. Figure 34 shows simply how this can be done by the use of pairs of gears and three shafts.

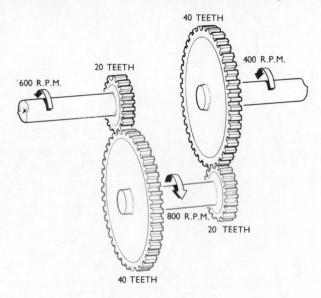

Fig. 34. A simple gear train.

This arrangement is almost the same as in a gearbox except that in an actual tractor gearbox, the operator can select one of a number of speed ratios by moving the appropriate gears into position. Figure 35 shows a possible arrangement in a gearbox which has three forward speeds and one reverse speed. This type of gearbox is known as a "sliding mesh gearbox" and is typical of the type used in tractors.

The clutch shaft, sometimes known as the *gearbox input shaft,* transmits the drive from the engine into the gearbox. The gear wheel on the end of the clutch shaft is fixed to it and in constant mesh with the gear on the bottom shaft which is known as the *layshaft.* All the gears on the layshaft are fixed to it, and as long as the drive from the clutch shaft is coming into the gearbox, the layshaft and all its gears will rotate.

The top shaft, known as the *output shaft,* carries the mating gears. These gears are splined on to this shaft and can be slid along it, but if they are rotated, then the shaft must also rotate and the drive is taken out to the rear wheels. Figure 35 shows the position of bottom gear (1st) with the gears *CC* in mesh. In this instance the drive is coming from gear *A* on the input shaft to *A* on the layshaft which is also turning gear *C.* This gear *C* on the layshaft, being in mesh with gear *C* on the output shaft, causes the shaft to rotate and the drive goes to the rear wheels.

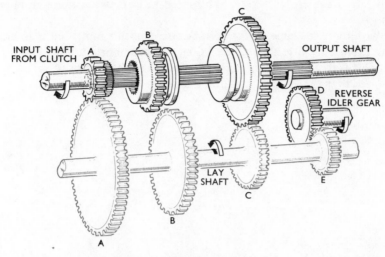

Fig. 35. A sliding mesh gearbox.

Second gear is provided by parting the *CC* gears and meshing the gears *BB*. The third gear is taken straight through from the input shaft by meshing together the gears *A* and *B* on the output shaft. This locks the output and input shafts as one.

When reverse gear is required, gear *C* is moved to mesh with gear *D* which is in constant mesh with gear *E*. By doing this, it can be seen that the direction of rotation of gear *C* will be reversed, therefore, the direction of rotation of the output shaft will also be reversed.

The tractor gearbox usually has more than three gear ratios, in fact as many as six or eight plus two reverse ratios may be used. To provide the additional gearbox range, a second system of gears in the same gearbox is used, which, when selected will give an alternative input speed. Such a gearbox would then have a high and low ratio.

A wide range of forward gears is needed because of the great variety of operations that a tractor must carry out; these operations vary from planting at perhaps less than 1 m.p.h. to road work at up to 25 m.p.h.

The type of gearbox used on tractors is more often than not the "sliding mesh" type, and because of this it is not usually possible to change gear when moving. If changing gear *is* attempted whilst moving, there is risk of causing damage to the gears involved. With such a gearbox it is usual to start off the tractor directly in the desired gear, and not to change up through the various gears as with a motor-car.

The Differential

So far, in the transmission system, the drive from the engine has come in a straight line with the engine crankshaft, but in order that the rear wheels of the tractor may be driven, it is now necessary for the drive to be taken at right angles to this line.

The gearbox output shaft extends a short distance outside the gearbox and has fitted to its end a toothed gear known as a *bevel pinion*. This pinion is in mesh with another bevel gear known as the *crown wheel*. Figure 36 shows how the drive changes direction. It should be noted that another speed reduction is brought about here by having a small pinion drive the large crown wheel. The speed ratio between these gears is usually about 5 to 1.

Whilst two objectives have so far been achieved in this unit, that is to change the direction of drive and cause a speed reduction, a third important function must also be achieved. This is to enable each rear driving wheel of the tractor to rotate at a faster or slower speed than the other when turning corners.

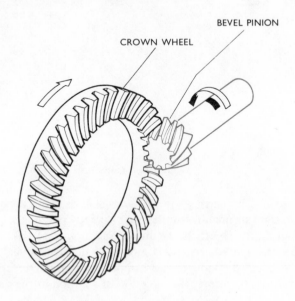

BEVEL PINION

CROWN WHEEL

Fig. 36. Crown wheel and pinion.

When a tractor is travelling in a straight line, both rear wheels rotate at the same speed, but when it turns, say, to the right, then the left or outside wheel must travel over a greater distance than the inside wheel. This means that the outside wheel must rotate faster than the inside wheel. If the tractor turns to the left, then again the outside wheel must rotate faster.

To enable this to be done, a unit known as the "differential" is used and it is directly attached to the crown wheel (see Fig. 37).

The axle shafts (half-shafts) which drive the rear wheels are not directly attached to the crown wheel but terminate with *bevel gears* which are in mesh with *differential pinions*. These bevel gears and differential pinions remain constantly in mesh. The pinions are attached to the crown wheel but are free to rotate on their spindles. However, when the crown wheel turns, they must turn with it.

When the tractor is being driven straight forward and there is an equal resistance applied to each wheel, the differential pinions will not rotate on their spindles but will be carried around with the crown wheel and will drive the two half-shafts in the same direction. However, when the resistance to the wheels is unequal, these pinions rotate on their spindles as well as driving round with the crown wheel. Referring to the illustration, supposing that the left-hand half-shaft is prevented from rotating, the right-hand half-shaft would still rotate, but because the differential pinions can rotate freely they will roll around the stationary bevel gear on the left-hand half-shaft.

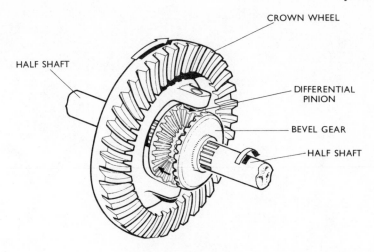

Fig. 37. The differential.

Of course, when the tractor is turning a corner, the inner wheel does not stop rotating entirely but slows down, and the outside wheel speeds up proportionately.

The fact that the differential unit will permit one wheel to rotate faster than the other is sometimes a disadvantage. For example, when a tractor is ploughing, one of its wheels is usually running on a greasy field surface. These conditions can immediately result in unequal resistance to the wheels, so that the wheel on the field surface having the least resistance spins around whilst the furrow wheel does not rotate.

To overcome this problem the modern tractor can be fitted with a *differential lock*. This device is usually foot-operated and can be brought into use when wheel spin takes place. Its purpose is to lock the differential unit so that in effect a solid back axle is formed and both rear wheels must drive together. When this lock is in use the tractor will travel in a straight line only, and it is not possible to turn corners.

Final Gear Reduction

All tractor gearboxes have some form of gear reduction as the drive leaves the differential unit before reaching the wheels. Two main types of gear reduction are used:

Spur reduction. This is where the half-shafts coming from the differential unit are fitted with a small spur gear coupled to a large spur gear wheel which is directly on the end of the rear axle (see Fig. 38).

Epicyclic reduction. This is where the drive from the differential enters an epicyclic gear pack and the drive is reduced to rear wheel speed. This unit gives the same drop in gear ratio as the spur reduction unit.

The Complete Transmission System

Throughout the transmission system the input speed of the engine crankshaft has been reduced in stages to give a suitable forward speed to the tractor. The gearbox of course, provides a number of alternative speeds.

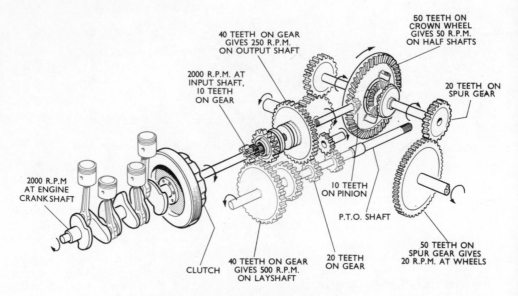

Fig. 38. The complete transmission system.

Figure 38 shows the complete layout of the transmission system. Shown in bottom gear with the engine crankshaft speed at 2000 r.p.m., the wheel speed is 20 r.p.m., thus giving an overall reduction of 100 to 1. Of course, whilst it provides a suitable forward speed by having a number of overall speed reductions in the transmission system, the transmission gearing also makes the power of the engine more easily applied to the work.

Wheels and Tyres

The rear wheels of the general-purpose tractor propel it forward, or backward, as the case may be, thus enabling the power of the engine to be used to move an implement or some other load. The front wheels are merely idler wheels by which the tractor can be steered in various directions.

Because of the many varying and adverse conditions in which tractors must work, their wheels must be equipped with tyres that will allow for the maximum possible grip. The tyres are therefore made large in diameter, wide, with heavy tread bars, and generally the casing is quite thin. The large diameter, the good width and the heavy tread bars provide a good area of tyre that can be in contact with the field surface.

For a given size of wheel fitted to a tractor it is generally possible to fit alternative sizes of tyres. Tyre sizes given by the makers refer to the width of the tyre as measured across from one wall side to another and the diameter across the inner rim of the tyre. A tyre therefore of 304 × 914 mm (12 × 36 in.) dimensions would be 304 mm (12 in.) wide and would fit a wheel rim which was 914 mm (36 in.) in diameter. Typical tyre sizes used on agricultural tractor rear wheels are 279 × 914 mm (11 × 36 in.), 304 × 914 mm (12 × 36 in.), 254 × 711 mm (10 × 28 in.), 279 × 711 mm (11 × 28 in.), whilst typical front wheel tyre sizes are 101 × 482 mm (4 × 19 in.), 152 × 482 mm (6 × 19 in.), 152 × 406 mm (6 × 16 in.).

The use of the wide tyres on the rear wheels will help improve grip because a larger area of tyre is in contact with the field surface.

Tyres are also made in different casing thicknesses and are given what is known as a ply rating. When tyres are made they are built up from a number of layers of fabric or canvas material on to which is moulded the rubber tread. The ply rating refers to the number of layers of canvas, etc., used in the tyre construction. Common thicknesses of casings used on agricultural tyres are four-ply and six-ply. The latter being thicker are stronger and will support more weight and be less affected by rippling.

These casing thicknesses are thin enough to allow the tyre to flex when riding over rough surfaces.

Even though tractor tyres are constructed as they are, conditions are still met where the wheels will tend to slip. The amount of grip that a tractor will be able to get will depend on the field conditions and these vary considerably. When a tractor wheel is turning, the tread bars of the tyre are tending to shear off layers of soil and it is the soil's resistance to this shearing action that helps the wheels to turn. This, of course, is just one factor that contributes to rotation of the wheels but is perhaps the most important. Obviously, the ease with which a layer of soil will shear off is going to affect the drive of the tractor wheels. Therefore in some soil conditions a lot of wheel slip may take place.

Methods of Reducing Wheel Slip

A typical tyre pressure for the rear wheels of a tractor is 1 kg/cm² (15 lb/in² approx.) when used for road work, and 0·7 kg/cm² (10 lb/in² approx.) when used for field work.

If the tyre pressure is reduced to, say, 0·5 kg/cm² (8 lb/in² approx.) for field work, the effect is to put a greater area of the tyre casing in contact with the field surface because the tyre flattens out. A greater area and number of the tyre tread bars will also come in contact with the field surface (see Fig. 39). This results in improved traction, but very low pressures can lead to damage to the tyres. The pressures should not be reduced to the point where the tyre wall can be seen to ripple as the tractor moves. This rippling breaks up the tyre casing and this can take place in a very short time, also there is a risk of the tyre creeping round the rim and pulling the valve out of the tube.

We do not worry so much about the wheel grip when a tractor is used for roadwork because the road surfaces are usually good and grip is not usually a problem. However, tyre wear must be considered because of the high cost of these tyres, therefore the pressures are kept higher.

The use of *wheel strakes* which bolt on to the wheel discs and protrude out to the perimeter of the tyre so that they bite into the earth, or the use of *wheel girdles* which fit entirely around the tyres are helpful in poor field conditions.

Perhaps one of the most used aids to wheel grip is that of additional weight in the form of either metal weights or water. The metal weights are in the form of round discs weighing up to 50 kg (1 cwt approx.) each and are fixed to the wheel discs. As many as three or four may be attached to each wheel.

When water is used it is put into the tubes of the tyres by either pump or gravity flow and using a special valve which lets air out as water goes in. The tyres may be filled completely, or more often 75% filling is carried out, and when this is done, air

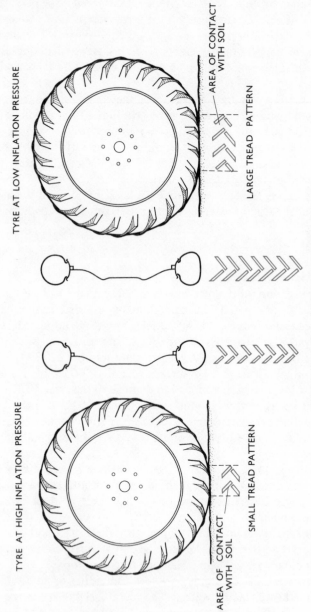

TYRE AT LOW INFLATION PRESSURE

AREA OF CONTACT WITH SOIL

LARGE TREAD PATTERN

TYRE AT HIGH INFLATION PRESSURE

AREA OF CONTACT WITH SOIL

SMALL TREAD PATTERN

Fig. 39.   The effect of altering tyre pressures.

must also be put in at about 0·2 kg/cm³ (3 lb/in² approx.) higher pressure than the normal used. This is done to help support the additional weight of water in the wheel which may be as much as 100-200 kg (2-4 cwt approx.) per wheel. A necessary precaution is to add anti-freeze to the water if it is intended to leave it in during the winter. Calcium chloride, used at a rate of 1 kg/4·5 litres (2·2 lb/gal) of water, provides sufficient protection against frost in this country.

Spade lug wheels are quite often used as a last resort when all other methods of improving traction have failed. These are metal wheels having around their perimeter steel lugs which bite into the earth to provide grip. When these are fitted to the tractor, the tractor must not be used on the public road.

Front Wheels

Because these are idler wheels supporting the front end of the tractor and allowing the tractor to be steered it does not mean that they should be neglected. Indeed, should the tyre pressure of these wheels vary, the tractor will become difficult to steer properly. It will tend to pull towards the side of the tractor which has the lower front wheel tyre pressure.

Tyre pressures for these wheels are usually about 1·75-2 kg/cm² (25-30 lb/in² approx.) but there are times when it may be necessary to increase the recommended pressure. Such an occasion would be when a fore-loader was being used on the tractor to lift heavy loads. Typical examples are loading manure with a front-mounted fork, loading soil with a bucket, or loading bales with a front-mounted bale loader.

It is usually possible to add wheel weights to the front wheels as well as the rear, and front wheel weights may be used when the tractor has to spend some considerable time carrying heavy rear-mounted implements. The front weights would then act as a counterbalance to prevent rearing up of the front of the tractor when an implement was lifted. Another method of applying front end weight is by fitting a special bracket to the front of the tractor to carry large specially shaped heavy weights which can be lifted on and off as required for any specific task.

Wheel Positions

On many farms a considerable amount of the tractor's working life is occupied doing rowcrop cultivations and because not all crops grown in rows are grown at a standard row width, it is necessary to be able to adjust the track width to suit the crop. For example, if potatoes are grown in rows 711 mm (28 in.) apart, then in order that the tractor wheels may run between these rows without damaging the crop, the wheels must be set at a multiple of 711 mm (28 in.). The wheel setting is usually 2 × 711 mm = 1 m 422 mm (2 × 28 in. = 56 in.).

Sugar beet grows in rows 457 mm (18 in.) apart, therefore the wheel positions could be set at 3 × 457 mm = 1 m 371 mm (3 × 18 in. = 54 in.) centres or 4 × 457 mm = 1 m 828 mm (4 × 18 in. = 72 in.) centres.

To make these alterations and settings possible for the front wheels of the tractor, the front axle is usually made in three pieces, each piece being made with an adequate number of holes to allow for wheel width alterations in 5-cm (2 in. approx.) steps (see Fig. 40a). It will be seen that one of the axle pieces is pivoted about the centre of the tractor, beneath the radiator. The narrowest wheel setting obtainable on a general

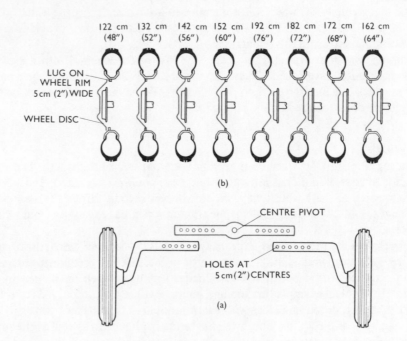

Figs. 40a and b. Wheel positions.

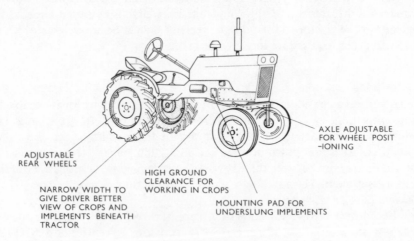

Fig. 41. A rowcrop tractor.

purpose tractor is usually about 1 m 219 mm (48 in.) whilst the widest may be as much as 2 m 32 mm (80 in.).

To achieve these same variations with the rear wheels, use is made of concave wheel discs which may be bolted in alternative positions to lugs welded on the wheel rim. These concave discs may also be used either with the concave facing inwards or outwards. Figure 40b shows the various settings of the rear wheels. Figure 41 shows a tractor designed mainly for rowcrop work.

Wheel Brakes

It is most important to have efficient brakes on an agricultural tractor and at no time should their maintenance and adjustment be neglected. Inefficient brakes have been the cause of many serious accidents on and about farms.

The type of brakes fitted to tractors are usually either *internal expanding brakes* or *disc brakes*. These two types are generally used, but a third type known as *external contracting brakes* is to be found on track-laying tractors.

All these brakes operate on the same principle in that a stationary component or components are mechanically moved so that they come in contact with a rotating component. The stationary components are lined with a hard-wearing friction material similar to that used for clutch discs, and when they are moved into contact with the rotating member, braking takes place.

Figure 42 shows an external contracting brake. The rotating member would be fixed to the drive shaft coming from the differential. Movement of the hand lever in the direction of the arrow will cause braking to take place because the brake band will tighten on the rotating drum.

And internal expanding brake is shown in Fig. 43. In this brake two brake shoes are moved outwards by a mechanism operated from the foot pedal. A brake drum is fixed to the rotating member and this drum totally encloses the brake's shoes. The

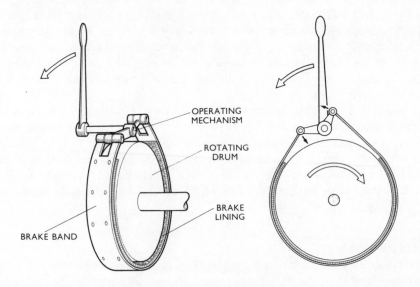

Fig. 42. The external contracting brake.

tractor wheel is also fitted to the rotating member so that when the tractor is moving, both drum and wheel are turning as one. When the brake pedal is depressed the partial rotation of the brake cam at the top of the back plate moves the brake shoes, which are pivoted at the bottom, outwards, so that they come into contact with the brake drum.

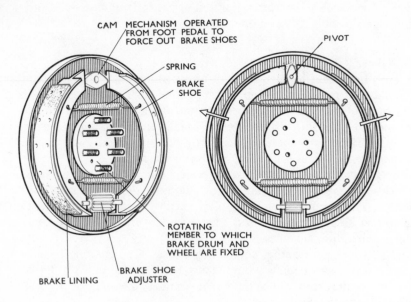

Fig. 43. The internal expanding brake.

Many tractors are now being fitted with disc brakes (Fig. 44). In this brake two rotating friction discs are used and these are caused to rotate because they are splined on to the differential shaft. Being on a splined shaft they will also move sideways if necessary. Also in this brake are two movable plates operated by the foot pedal, and two fixed plates. These latter are actually the inner sides of the brake housing. The operation of the brake consists of forcing the two movable plates outwards so that the friction discs are prevented from rotating. This will stop rotation of the differential shaft which is driving the tractor wheels.

The position at which the brakes are fitted can vary. On tractors that do not have a spur reduction to the rear wheels, the brakes must be fitted to brake the shaft to which the wheel is attached. However, if a spur reduction *is* used, it is usual to fit the brakes to the first shaft coming from the differential. These shafts can be seen in Fig. 38.

One of the advantages of having the brakes in this position is that they are kept well out of the way of mud and the like which may get into the brake drums to cause faulty braking.

Independent Braking

When tractors are used for such field work as ploughing or rowcrop cultivations, it is important that they should be able to turn sharply into the bouts of work because usually the area in which turning must take place on a headland is restricted.

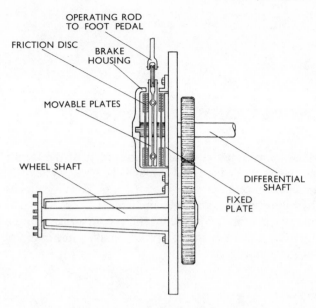

Fig. 44. The disc brake.

To assist in this, the brakes of tractors can normally be operated independently of each other. If the right-wheel brake is applied, the tractor will tend to pivot on this wheel and rotation of the left wheel will cause it to turn sharply to the right.

The brakes are operated by foot pedals and while these pedals can be operated independently of each other, they can also be latched together so that both brakes operate at one time if necessary. This, of course, would have to be the case when the tractor is used for road work.

A parking brake is also required on a tractor and this is usually a ratchet device which holds the pedals down in a locked position.

CHAPTER 7

The Tractor Hydraulic System

NO MECHANISM has contributed more to the usefulness of the farm tractor than has the hydraulic mechanism. Virtually all tractors have this system fitted as standard equipment with the exception of crawler-type tractors; however these tractors can also be fitted with a hydraulic system as extra equipment.

Its main use is in the operating of field machinery for either above or below surface implements. It can also provide power to operate remote rams or motors on harvesting, hedging or ditching machinery, etc.

Operating Principles

In a simple hydraulic system as shown in Figs. 45 a and b, the pressure is the same everywhere in the system. The illustration shows two cylinders connected together by a pipe and the cylinders and pipe contain oil. In the small cylinder is a neatly fitting piston which has a surface area of 1 cm² or (0·155 in²), in the larger cylinder there is a ram piston with a surface area of 10 cm² (1·55 in²).

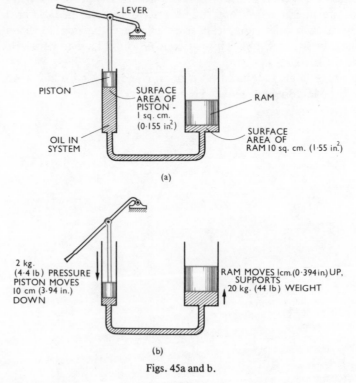

Figs. 45a and b.

Now if a pressure of, say, 2 kg (4·4 lb) is applied to the small piston, that pressure is being applied to 1 cm² surface area of the oil. Because the pressure is the same everywhere in the system we find that for every square centimetre of surface area beneath the ram piston there is also 2 kg (4·4 lb) pressure. Because the area of the ram piston face is 10 cm² (1·55 in²), the ram will support a weight of 2 × 10 = 20 kg (4·4 × 10 = 44 lb). So by applying a small weight at one point we can support a large weight at another point. In this case the larger weight is 10 times greater than the smaller weight. In doing this to move the large weight 1 cm (0·394 in.), the small weight must move 10 cm (3·94 in.). The ratio of movement is the same as the ratio between the weight. Likewise it is the same as the ratio between the cross-sectional areas of the piston and the ram piston.

If this simple system had in it a non-return valve to prevent the oil from being forced back by the weight each time the lever was lifted, and also had a sump supplying oil, it would be possible to lift a weight progressively higher by applying a pumping action to the piston. We would have a simple hydraulic jack.

In the tractor hydraulic system the pump supplying oil to the ram cylinder is usually a gear pump driven by the engine or some part of the transmission system and pressures of up to 910 kg (2000 lb approx.) are provided.

It is necessary in such systems to have control over the flow of oil so that excessive pressures are not built up, therefore safety valves are incorporated as protection. The tractor driver operates the raising and lowering of an implement by the control lever fitted near his driving seat. Movement of the lever operates a valve which controls the flow of oil to the ram cylinder.

Figure 46 shows a basic tractor hydraulic system. The illustration shows the principles only, but the arrangement of ram cylinder, lift arms and linkage arms is typical. The intake of oil to the pump is filtered, usually by both gauze filter and magnetic filter. This is most important because the smallest particle of dirt or metal can cause considerable damage to the essential parts of the system. When changing oils in the hydraulic system, absolute cleanliness should be aimed at.

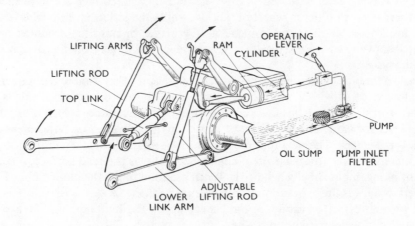

Fig. 46. A basic hydraulic lift and linkage arrangement.

The Three-point Linkage

The three-point linkage is shown in Fig. 46 and it consists of two lower links pivoted beneath the tractor's back axle, and a top link pivoted on the top of the back axle housing. The lower links are attached to the lifting arms by lifting rods and the right one of these is adjustable.

Figure 46 also shows how lifting and lowering of the lower linkage arms is brought about. Oil forced into the cylinder pushes back the ram which in turn causes the lift arms to move up. If the oil can be trapped in the cylinder, it will support a weight hung on the linkage arms, and lowering of the weight can be brought about by allowing the oil to escape from the cylinder when required. This is all achieved by having valves in the system which may be arranged to operate automatically or by hand lever control.

The ends of the linkage arms and top link are fitted with ball joints so that they will attach to suitable pins fitted on to an implement. These ball joints or pins are either category 1 which is 22 mm (7/8 in.) diameter or category 2 which is 28 mm (1 1/8 in.) diameter.

Fitting of the implement to the tractor entails reversing the tractor to the implement with the lower link arms down, and stopping when the ends of the arms come in line with the attachment pins on the implement. The non-adjustable (left) arm is fitted first to the implement and then secured, then the other arm which can be adjusted if required to bring it in line for attachment. Finally, the top link is attached to tractor and implement.

With an implement attached to a tractor in this way, they together form a single working unit. The implement is easily transported from one place to another and just as easily lifted in and out of work by the lift control lever which is conveniently placed for the operator to use. The method of operating in the field varies according to the design of the hydraulic system.

Field Operation

Implements for operation on the hydraulic lift system of tractors may be divided into two main classes: semi-mounted or fully mounted. The former are implements to which there is fitted a depth wheel or wheels, whilst the latter have no wheels. Both classes are known as mounted implements. Whether or not semi-mounted or fully mounted implements are used with a tractor depends largely on the design of the hydraulic system.

A basic hydraulic system will allow for the operation of semi-mounted implements, say a plough, fitted with a depth-control wheel. The field operation of this is fairly straightforward; the plough is lowered into work by the operator moving the hydraulic control lever forward, and the depth of ploughing is controlled by adjustment of the plough depth wheel. The weight of the implement is on the field and not on the tractor, which merely pulls it. Because the plough wheel follows the field surface, a constant depth of ploughing is usually achieved. When required to lift the plough out of work, the operator moves the hydraulic control lever rearward.

One obvious effect of lifting an implement up on the hydraulic lift linkage is the transference of weight to the rear wheels of the tractor. This is due to the weight of the implement itself and to its position of overhang on the linkage arms. This weight transference is used in some hydraulic system arrangements to give better wheel adhesion (traction).

Draught control is the name given to such systems that utilize weight transference, and in this the implements used are not fitted with depth-control wheels. This means that the weight of the implement is always carried on the linkage arms of the tractor so that there is considerable weight increase to the rear wheels of the tractor. Furthermore, when an implement is being worked in the soil there is an additional weight provided, due to the action of the soil on the implement.

In a system of this type, the working depth of the implement and also the maximum draught is controlled by the position of the hydraulic control lever which is operated by the tractor driver. When this is set to produce a desired depth of work, the draught is automatically maintained.

If a plough is attached to the lower link arms of the tractor and the top link is not fitted (see Fig. 47), then, when the tractor is moved forward, the point of the plough will dig into the ground. This will cause the plough to rotate about the points where it is attached to the link bars. The plough attachment point for the top link will move in a forward direction towards the rear of the tractor. Now even with the top link fitted to plough and tractor, and with the plough in work, the tendency for the plough to rotate is still there. This means that there is always a forward thrust through the top link to the point where it is attached to the tractor.

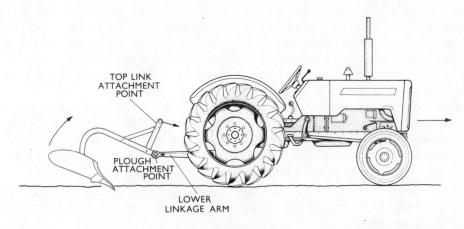

TOP LINK
ATTACHMENT
POINT

PLOUGH
ATTACHMENT
POINT

LOWER
LINKAGE ARM

Fig. 47.

The usual arrangement to maintain a constant draught is to have the end of the top link connected to the tractor in such a way that the forward thrust is transmitted to the mechanism which operates the hydraulic control valve. This is usually done through a heavy compression spring and before the force can be transmitted to the control valve it must reach a certain figure. Let us assume that this figure is 910 kg (2000 lb approx.). Then any forward thrust exceeding this will cause the mechanism to operate and the implement will lift until the thrust is again adjusted to 910 kg (2000 lb approx.).

When a plough is working, the force tending to cause it to rotate can vary considerably from one end of a field to the other. One of the main causes of it varying is the soil itself, simply because as soil texture varies, so will its resistance to being turned over by a plough body. Clay soil offers much more resistance than a loam soil. This, of course, means that the forward thrust in the top link will also vary according to soil conditions, with the result that the hydraulic mechanism automatically makes the necessary adjustment to maintain a constant draught.

In conditions where much difference in soil conditions occurred, the earlier type of draught-control system produced work of uneven depth, but the modern improved systems have eliminated much of this.

Some systems provide for an alteration in the speed at which the hydraulics respond to different soil conditions and the rate at which implements can be lifted and lowered. Many allow for the operation of both wheeled and wheel-less implements so that advantage can be taken of improved traction if required.

Another system provides what is known as "traction control" and the mechanism which brings it into operation can be set to provide increased weight on the rear of the tractor as and when required. The amount of increased weight that can be applied can be varied from about 100-250 kg (2-5 cwt approx.). The principle on which it operates is based on the fact that in attempting to lift a weight, before it ever moves, some of its weight is transferred to whoever or whatever is trying to lift it. For example, if a man tries to lift a weight of, say, 100 kg (2 cwt approx.), as he gradually increases his effort to lift it, so will the weight transferred to him gradually increase. This means that the pressure exerted by his feet on the ground will gradually increase until, when he lifts the weight, the ground pressure will be the man's weight plus 100 kg (2 cwt approx.). The traction control unit is designed so that the hydraulics only attempt to lift the weight of the implement, and back axle weight is increased without any actual lifting of the implement taking place.

Very large modern tractors now operate ploughs which due to their construction would not give any top link sensing action for draught control. Therefore other methods of sensing have been adopted on large tractors.

Lower link sensing. This system is operated by the pull of the implement against the lower links. This pull operates a cranked arm on the lower link front hitch point which is spring loaded by either conventional springs or torsion bars. The amount of pull operates an hydraulic spool valve which in turn lifts or lowers the implement slightly to maintain a constant draught, see Fig. 48.

Load monitor. This employs a small compact unit inserted in the drive shaft between the gearbox and diff unit. It splits the drive and is loaded by a heavy spring.

When the tractor meets a heavy draught load the unit is forced apart by cam action against spring pressure. This movement works a forked lever which is connected to the hydraulic valve chest, and operates the draught sensing similar to the other methods previously described.

Auxiliary Hydraulic Tappings

The tractor's hydraulic system also provides a useful source of hydraulic power for the operation of implements or parts of implements remotely connected to the tractor. Typical examples of the use of this power can be found on hydraulic tipping trailers, fore-end loaders, rear loaders, ditching machinery and driving hydraulic motors.

Nearly all modern tractors are provided with outlet points where flexible pipes can be fitted to convey oil from the hydraulic pump to an external ram and cylinder fitted to an implement. Oil, pressure fed to the ram, can then be used to raise or move parts of implements. For example, in the tipping trailer a flexible pipe is connected between a self-sealing outlet union on the tractor, to a ram and cylinder beneath the trailer body (see Fig. 49). The trailer body can then be raised by operation of the hydraulic control lever. Lowering is brought about by moving the control valve so that the weight of the trailer body forces the oil out of the cylinder.

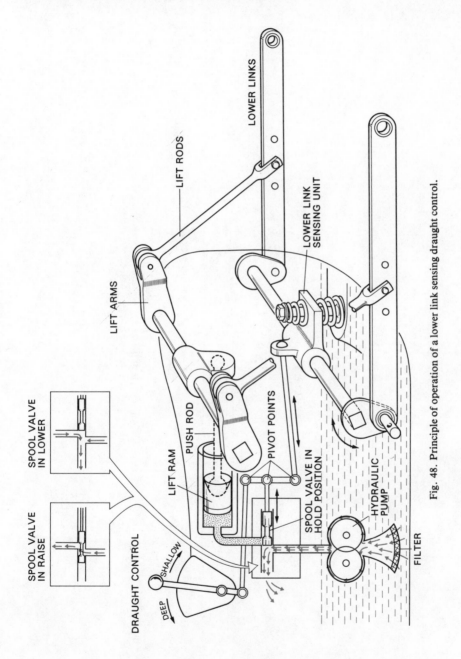

Fig. 48. Principle of operation of a lower link sensing draught control.

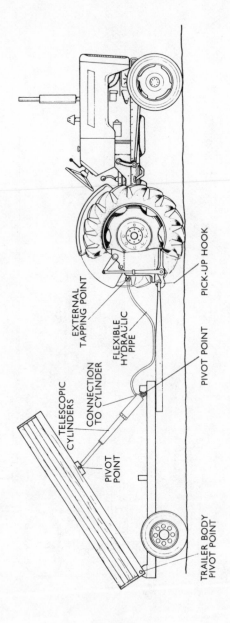

EXTERNAL TAPPING POINT

FLEXIBLE HYDRAULIC PIPE

PICK-UP HOOK

PIVOT POINT

TELESCOPIC CYLINDERS

CONNECTION TO CYLINDER

PIVOT POINT

TRAILER BODY PIVOT POINT

Fig. 49. An example of the tractor's hydraulic system being used to operate a tipping trailer.

CHAPTER 8

The Power Take-off Shaft

THE Power Take-off Shaft, usually called the p.t.o., provides us with a means of driving machines that are being towed by the tractor. The engine power of the tractor is not only being used to move the tractor and pull or carry the machine, but also used to operate the machine. Typical examples of machines that are power driven are mowers, balers, combines, potato harvesters and manure spreaders. This provides a very useful source of power and the alternative would be to fit an engine to each of these machines if no p.t.o. was available. This is, of course, done where the power required to drive the machinery may be greater than the tractor can supply. Generally though, most farm machines can be powered by the tractor.

The p.t.o. shaft is an extension of the gearbox layshaft which is being continuously driven (see Fig. 38). The only way to prevent the layshaft from rotating is to disconnect the engine clutch, and this, of course, will also stop rotation of the p.t.o. shaft.

This is not a satisfactory arrangement simply because if a tractor driver has, for some reason or other, to get off his tractor to make some adjustment to the machine being driven, the p.t.o. would continue to drive as soon as he takes his foot off the clutch pedal. Furthermore, the machine would also be driving whilst being transported from farm to field or field to field provided that the drive shafts were coupled together.

It is therefore necessary to provide a means whereby the p.t.o. shaft can be disengaged independent of the engine and transmission. This is done simply by putting a clutch device in the p.t.o. line and arranging it so that it can be engaged and disengaged by a lever positioned within convenient reach of the tractor driver. Figure 50 shows an alternative arrangement where two separate shafts are used.

The end of the p.t.o. shaft protrudes outside the rear end of the tractor for a distance of about 10 cm (4 in. approx.). This end is splined so that a suitable coupling, internally splined, can be slid on to it and locked in position to form a solid drive (Fig. 50).

The type of coupling shaft used to transmit the power of the engine to the machine being driven is fitted with two *universal joints* and it is made so that it is telescopic and always rotates even when the machine is not in a straight line behind the tractor.

An open or uncovered drive shaft from a tractor p.t.o. to a machine being driven is a most dangerous unit. To work a machine without having this shaft suitably guarded is foolish practice and is illegal. Many a tractor driver has been severely injured because loose clothing which he was wearing has caught up in the shaft.

There are two sizes and types of p.t.o. shafts.

The standard 6-spline 35 mm (1 3/8 in.) operating at 540 r.p.m.

The new 21-spline 35 mm (1 3/8 in.) diameter operating at 1000 r.p.m.

Both of these shafts rotate in clockwise direction when viewed from the rear of the tractor.

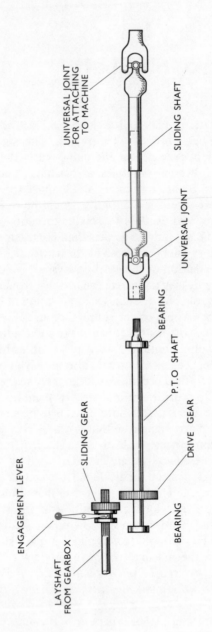

Fig. 50. A power take-off shaft and universal joints.

Most tractors can be fitted with both the above types of p.t.o. shafts, or with interchangeable shafts which automatically alter the output speed when engaged and fitted.

The Live P.t.o.

The standard p.t.o. previously described will stop rotating if the tractor engine clutch is disengaged by depressing the clutch pedal. This is because the clutch disconnects the engine from the gearbox, which drives the p.t.o. shaft. Forward movement of the tractor will be also stopped.

In some field operations where power-driven implements are used this arrangement can be a handicap, because it is often necessary to stop the forward movement of the tractor without stopping the p.t.o. shaft from driving the implement; for example, when baling hay it is often necessary to allow the pick-up to clear itself of hay before more is taken in. To overcome this the *live p.t.o.* was introduced.

One arrangement to provide a live p.t.o. shaft is shown diagrammatically in Fig. 51. A dual clutch is used and it incorporates two clutch discs, one of which transmits the drive from the engine to the transmission and the other transmits the drive to the p.t.o. shaft. This is done through the medium of hollow and solid drive shafts. The arrows shown on the diagram indicate the path of the drive to the p.t.o. shaft.

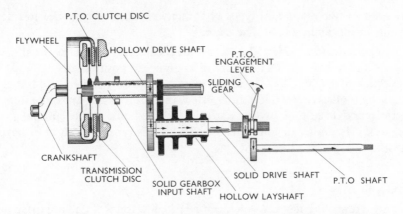

Fig. 51. An arrangement providing a "live" power take-off shaft.

The p.t.o. clutch disc is splined on to the hollow drive shaft to which is fitted a drive gear. This gear is in mesh with another that is fitted to a solid shaft that can rotate within a hollow gearbox layshaft. The end of the solid shaft is shown splined and carries a sliding gear which can be meshed with the fixed gear on the p.t.o. shaft when the engagement lever is moved in the appropriate direction. This means that as long as the p.t.o. clutch disc is gripped between the clutch pressure plates the drive will go through to the p.t.o. shaft.

The mechanical arrangement of the dual clutch is such that it provides two-stage operation. (For the purpose of clarity the mechanical detail of the operation of the clutch has been left out of Fig. 51). When the clutch pedal is depressed through about half its length of travel, the transmission clutch is released and this stops the forward movement of the tractor. However, it is not until the pedal is fully depressed that the p.t.o. clutch is disengaged, and the p.t.o. shaft will stop rotating. This gives us an effective means of stopping the tractor from moving forward whilst still allowing the power shaft to drive an implement. Fig. 33 page 51 shows a section through a dual clutch. The p.t.o. engagement lever is still needed as there must be some means of disengaging the drive entirely when required.

On many farms there is a need to drive certain machines by a flat belt, e.g. saw benches, hammer mills, emergency generators, irrigation pumps, crop-drying fans, etc.

Most tractors can be equipped with a belt pulley for driving the above equipment. The pulley unit usually fits over the p.t.o. shaft at the rear of the tractor and uses the normal p.t.o. gear levers and clutch.

The tachometer usually displays standard belt pulley speed.

Guards must be fitted to comply with Ministry standards.

For information regarding pulleys and calculations of speeds and pulley sizes, see Chapter 9.

Belt drives, however driven, are dangerous units and should always be treated with respect. They *must always* be suitably guarded to prevent accidents. *Never* knock belts off pulleys while they are running.

Hitches

There are two fairly standard types of fitting which can be attached to tractors for towing implements and trailers. These are:

The swinging drawbar (Fig. 52a). It is attached to the underside of the rear axle housing and can be offset from its centre to a number of positions.

The pick-up hitch (Fig. 52b). It is used mainly for towing trailers and is operated by the tractor hydraulics. When it is fully lifted, the hook comes up to the underside of the tractor rear axle so that anything attached to the hook cannot normally come off (see Fig. 49). The trailer or implement used with this type of hitch must have a ring-type drawbar.

Other Devices

There are a few other devices which can be fitted to tractors and are often necessary for the correct operation of certain machines and implements. These are:

Check chains (Fig. 52c). Their purpose is to prevent the linkage arms from swinging out and striking the tractor wheels. They are often adjustable in length so that the side to side swing of certain implements attached to the linkage arms can be limited if necessary.

Stabilizers. These are also used to prevent side to side movement of certain farm implements. They usually take the form of metal bars slightly cranked at each end and they fit between an anchorage point beneath the rear axle and the ends of the linkage bars. Sometimes they are made adjustable in length as shown in Fig. 52d.

Top link. This has been mentioned in the section on three-point linkage. It is used in conjunction with the side linkage bars at the rear of the tractor to carry a rear-mounted implement. Figure 52e shows an adjustable top link. The adjustment is used to alter the pitch of an implement.

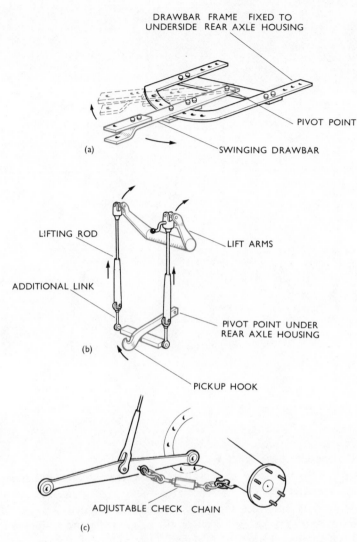

DRAWBAR FRAME FIXED TO
UNDERSIDE REAR AXLE HOUSING

PIVOT POINT

(a)   SWINGING DRAWBAR

LIFTING ROD

LIFT ARMS

ADDITIONAL LINK

PIVOT POINT UNDER
REAR AXLE HOUSING

(b)

PICKUP HOOK

ADJUSTABLE CHECK CHAIN

(c)

Figs. 52a, b and c.

Drawbar pin. This is used to couple an implement or trailer to the tractor drawbar. A good drawbar pin should be made of steel, be no less than 19 mm (3/4 in.) diameter and have a hole drilled in the end for fitting of a *linch pin* (Fig. 52f). The linch pin is essential to prevent the drawbar pin from jumping out whilst a load is being hauled.

The failure of a drawbar pin to support the weight of a loaded trailer or heavy machine has been the cause of many accidents on farms and roadways. It is not uncommon to see heavy loads attached to the drawbar of a tractor by a pin totally unsuitable for the job it has to do. Such things as 13-mm (1/2 in.) bolts are quite often used and these invariably bend or pull out after very little use.

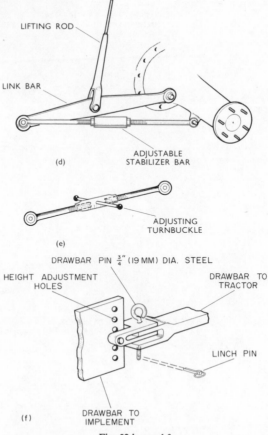

Fig. 52d, e and f.

Routine Maintenance of the Tractor

If a tractor is to serve its owner well, it must receive regular maintenance attention. If it does not receive this attention, stoppages due to mechanical breakdown are likely to occur. All tractor manufacturers provide excellent instruction books for use by the tractor operator and in these books full details of servicing and maintenance are given. The form in which this information is given is generally such that servicing attention to the various parts of the tractor is recommended to be carried out at intervals during the working life of the tractor. Some items may require attention daily when the tractor is in use, others every so many hours of tractor use.

If this servicing procedure is to be carried out on the basis of hours worked, etc., some record of the hours worked by the tractor ought to be kept. A tractor log book in which the operator can record work times is useful for this. The type of servicing and maintenance done, quantities of fuel and oils used, etc., can also be recorded in the log.

Most tractors can be fitted with an instrument called a tachometer or hourmeter which records the engine speed, p.t.o. speed, belt pulley speed, ground speed (k.p.h. or m.p.h.) and the engine operating hours. The latter is useful because it tells the operator how many hours the engine has been running and he can then do the necessary servicing according to that which is recommended.

A typical servicing and maintenance scheme for a diesel tractor would be as follows:

Daily attention

    (a) Check the oil level in the engine sump and fill up to full mark on the dipstick if necessary.

    (b) Check the water level in the radiator and fill up if necessary.

    (c) Check oil level in the oil bath air cleaner and clean out dirt if necessary. Remove any heavy trash from the filter screens.

    (d) Lubricate steering linkages with grease gun.

Every 50 hours

    (a) Remove the air cleaner oil container and also the filter. Wash out the container and refill with engine oil. Wash the filter screen thoroughly in paraffin, then blow out with an air-line. (Replacing a filter screen that is soaked in paraffin can cause a sudden build up of engine revs on starting the engine.)

    (b) Check the oil level in the gearbox and fill up to the full mark on the dipstick if necessary.

    (c) Check the oil level in the rear axle and fill up to full mark on the dipstick if necessary.

    (d) Lubricate front and rear wheel bearings.

    (e) Lubricate clutch cross-shaft.

    (f) Lubricate brake pedal shaft.

    (g) Check tyres and inflate to correct pressures.

    (h) Top up battery with distilled water and clean the battery terminals.

Every 200 hours

    (a) Remove and clean engine air breather filter, lightly oil the filter mesh. .

    (b) Drain the engine lubricating oil and refill the sump with new oil.

    (c) Remove the engine external oil filter element, wash out the bowl and refit a new element.

    (d) Lubricate the generator.

    (e) Check the oil level in the steering box.

Every 600 hours

    (a) Remove the diesel fuel filter element, wash out the bowl and fit a new element.

    (b) Remove the fuel injectors and replace with a service set.

Every 1000 hours

    (a) Drain the gearbox lubricating oil and refill the sump with new oil.

    (b) Drain the rear axle lubricating oil. Remove and clean the hydraulic pump filter screen and magnetic filters. Replace filters and refill the sump with new oil. Drain and flush the cooling system.

The items mentioned in the servicing and maintenance scheme above are nearly all concerned with lubrication. There are other necessary points of attention required in the use of a tractor. For example, brakes will require adjusting periodically, the clutch-pedal movement and the fan-belt tension will require adjusting also. These adjustments are made as required, and detailed information on how to make them will be found in the operator's instructions applicable to the tractor.

A good tractor operator will also keep his tractor clean and will occasionally check the tightness of nuts and bolts that may have loosened due to vibration.

CHAPTER 9

Mechanical Principles

Force

Force is anything which changes or tends to change the state of rest, or uniform motion, of a body. Thus a tractor pulling an implement exerts a force on the implement. The latter exerts a force on the soil or crop. A sack of corn on the ground exerts a force on the ground.

Under the SI (Systéme International) system of units force is measured in newtons (N) which may be defined as follows:

"When a force of one newton is applied to a mass of one kilogram (kg) it will give it an acceleration of one metre per second per second".

Mass

This is measured in units of kilograms and defines the quantity of material a body contains, for example a 50-kg bag of fertilizer. Knowledge of the mass of a body is useful when one is trying to determine application on spreading rates, e.g. 175 kg of fertilizer to be spread over 1 hectare (175 kg per ha). Using this situation it can be seen that mass is indeed different to force because our bag of fertilizer only exerts a force when we have to overcome it. If we try to lift the fertilizer, we have to overcome a force of 500 N in order for the bag to move. Unfortunately force and mass are frequently mis-used.

The attraction by gravity on a mass of 1 kg is 9·81 N (approximated to 10 newtons), thus to be able to appreciate the magnitude of a newton it is approximately the same as the force a 100-gram weight exerts upon your hand.

Pressure

Pressure is applied when a force (N) is applied to a given area, e.g. square metre (m^2), and is therefore measured in N/m^2. Thus when we look at an object, be it light or heavy, it will be the area of surface contact that will determine the pressure.

Example:

A caterpillar tractor has a mass of 10 tonnes (10,000 kg) over a total track area of 12 m^2. What is the pressure on the soil?

$$10,000 \, kg \times 9·81 = 98,100 \, N$$
$$\frac{98,100}{12} = 8175 \, N/m^2 \text{ or } 8·175 \, kN/m^2.$$

Example:

A push bike and rider has a mass of 80 kg and the total tyre contact with the ground equals 0·0025 m². What is the pressure exerted on the ground?

$$80 \text{ kg} \times 9 \cdot 81 = 784 \cdot 8 \text{ N}$$

$$\frac{784 \cdot 8}{0 \cdot 0025} = 313{,}920 \text{ N/m}^2 \text{ or } 313 \cdot 9 \text{ kN/m}^2.$$

As can be seen from the above examples, the considerably heavier vehicle actually exerts less ground pressure than the lighter vehicle, thus making it capable of travelling over much softer ground without sinking in.

It should of course also be understood that a force can be applied in a number of ways. Force can be applied to an object in four ways: tension, compression, torsion and shear.

(a) The object may be pulled, when the force is said to be in tension; e.g. a tractor pulling a trailer or the load of some other farm implement.

(b) The object may be pushed, when the force is said to be in compression; e.g. a tractor pushing a fore-end loader into a heap of manure.

(c) The object may be twisted, when the force is said to be in torsion; e.g. the rear axle shaft of a tractor twists the rear wheels, so that the tractor travels along the ground.

(d) The object may tend to be cut or sliced, when the force is said to be acting in shear; e.g. a mower blade traps the grass between the knife section and ledger plate when the shearing force is large enough to cut through or "shear" the grass.

Figure 53 shows the effects of some of the forces. The trailer drawbar is in tension due to the tractor pulling the trailer. The ground is in compression due to the weight of the tractor and trailer. The rear axle shafts of the tractor are in torsion, and the drawbar pin is in shear due to the pull on the tractor and the load imposed by the trailer. There are also a great many other very complicated forces acting on the tractor and trailer.

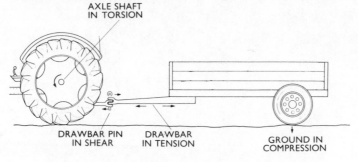

Fig. 53. Some of the forces acting on a tractor and trailer.

Work

When a force is applied to a body causing it to move through a distance, work is said to be done. The amount of work is the multiple of force × distance.

Thus: Work = Force (newtons) × Distance (metres)

Work = Nm.

The unit of 1 Nm may also be called a joule (J).

Power

This is a measurement of the rate of doing work and is derived by dividing the work done by the time taken.

$$1 \text{ joule/second} = 1 \text{ watt}.$$

Drawbar Power

This is often quoted by tractor manufacturers and is a measurement of the tractors load pulling ability — usually undertaken on concrete to minimize wheel slip.

Example:

A tractor travelling at 6 kilometers per hour exerts a pull of 20 kN. What is the drawbar power being developed?

$$6 \text{ km/hr} = \frac{6000}{60 \times 60} = 1.33 \text{ m/s}.$$

Work done per second $\therefore$
$$\begin{aligned} &= 20{,}000 \times 1.33 \\ &= 26{,}600 \text{ Nm/s}. \\ &= 26.6 \text{ kNm/s}. \end{aligned}$$

$$\text{Drawbar Power} = 26.6 \text{ kW}.$$

Brake Power

This is determined by measuring the turning force of an engine at the flywheel and may be undertaken by using a special machine called a dynamometer.

The turning force is referred to as the TORQUE of an engine and again is measured in newton metres (Nm). See below.

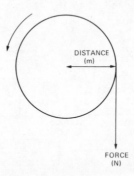

Measurement of torque.

Because a flywheel rotates, its speed which is normally measured in revolutions per minute (rpm) must be converted to angular velocity.

$$\text{Angular velocity} = \text{Speed (rpm)} \times 2\pi.$$

Example:
A diesel engine tractor is developing a torque of 250 Nm at an engine speed of 1200 rpm. What power is being developed?

$$\text{Brake power} = \frac{\text{Torque} \times 2\pi \times \text{rpm}}{60}$$

$$= \frac{250 \times 2\pi \times 1200}{60}$$

$$= 250 \times 2\pi \times 20$$

$$= 31420 \text{ watts}$$

$$= 31 \cdot 42 \text{ kW.}$$

Example:
An elevator lifts 6000 kg of grain per hour to a height of 10 metres. What power is being developed?

$$6000 \text{ kg mass} = 6000 \times 10 \text{ newtons (approx.)}$$
$$\text{Force} = 60{,}000 \text{ Nm.}$$
$$\text{Work done per hour} = 60{,}000 \text{ Nm}$$
$$\text{Work done per second} = \frac{60{,}000}{60 \times 60} = 166 \cdot 7 \text{ Nm/S.}$$

$$\text{Power} = 166 \cdot 7 \text{ watts}$$

Levers
Every day everyone of us uses a lever. It may be a coin to remove a clip-on top from a tin, a brake lever or a wheel barrow. The lever is a device for increasing the force which we can exert, for increasing the length of movement which we wish to make or changing the direction of a force.

Principle of Levers
Figure 54A represents a lever. A beam rests on a pivot P; there is a mass M resting on the beam at point X. To balance the beam we can push downwards at Y with an equal force to that exerted by the mass M. As long as the distance PX is equal to PY.

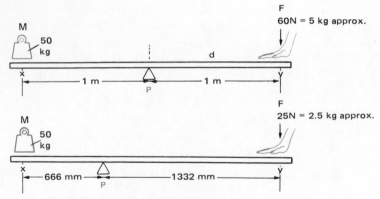

Figs. 54a and b. Principle of the lever.

But supposing we push downwards with a force equal to only half the mass *M*. Looking at Fig. 54b it will be seen that we apply this force at a point which is twice as far from the pivot point in order to obtain balance. Now the mass *M* is being held by a force which is only half as great. It would seem that we have got "something for nothing", but this is not true. If we wanted to lift the mass *M* a distance of 1 metre we would have to push down at *Y* a distance of 2 metres, thus the work (force × distance) being done is the same at both ends of the beam.

So it will be seen that the effect of a force depends not only on how large it is but also on where it is applied, and that if we wish to lift a load which is larger than the force which we can exert, the force must be moved through a greater distance than the load will be lifted.

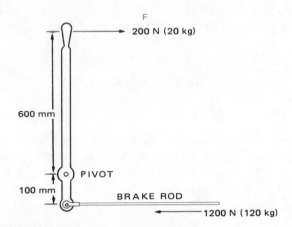

Fig. 55. Forces in a brake lever.

Figure 55 shows a brake lever. By pulling on the handle of the lever we can operate the brakes with a force greater than that which we can apply at the handle end; but the movement on the handle will be greater than that on the brake rod.

Figure 56a shows a wheelbarrow. It is relatively easy to lift, say, a 50-kg sack of cement; because the handles are farther from the pivot, in this case the wheel, than is the load.

Figure 56b shows a hay fork lifting a sheaf. In this case, although the sheaf weighs, say, 5 kg, we have to exert a force greater than 50 N (5 kg) because we are lifting at a point perhaps only half-way along the fork shaft, while the sheaf is on the tines. The end of the handle is acting as the pivot. However, the sheaf is moved over a greater distance than the force which is used for lifting.

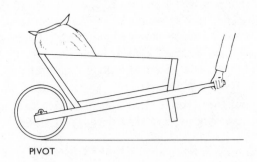

PIVOT

Fig. 56a. In the case of a wheelbarrow the handles are farther from the pivot than the wheels.

Fig. 56b. When lifting a sheaf with a fork, the end of the handle is the pivot.

Principle of Moments

It will be seen from the above that the leverage of a force depends not only on how strong it is but also how far it is from the pivot point. This leverage is called the *moment* of the force, and it is measured by multiplying together the strength of the force and its distance at right angles from the pivot. The moment of the force shown in Fig. 54a is $F \times d$, i.e. *Fd*. The units are similar to those used for the force and the distance; if the force is measured in newtons and the distance in metres, the moment of force is *Fd* newton metres (Nm) if the force is measured in pounds and the distance in feet, the moment of force is *Fd* foot-pounds.

If a lever balances, that is it is stationary, the moments of all the forces tending to turn the lever clockwise equal the moments of all the forces tending to turn the lever anticlockwise. Look at Fig. 55 which shows a hand brake. Let us work out the force F on the brake rod when a force of 200 N is applied to the hand brake at a distance of 600 mm (2 ft approx.) from the pivot, the brake rod being attached at 100 mm (4 in. approx.) from the pivot.

Taking moments about the pivot P, the moment of the force applied by the hand which tends to turn the lever clockwise $= 200 \times 0.6$ Nm $= 120$ Nm.

The moment of the force applied to the brake which tends to turn the lever anticlockwise and balance the forces, will be $100\,\text{mm} \times X$. Thus X can be calculated as follows:

$$200 \times 0.6\,\text{Nm} = X \times 0.1\,\text{Nm}$$

$$= 120\,\text{Nm} = 0.1X$$

$$\therefore X = \frac{120\,\text{Nm}}{0.1\,\text{m}} \quad \therefore X = 1200\,\text{N}.$$

It does not matter how many forces are applied; if the lever is in balance, the sum of all moments tending to turn the lever clockwise will be equal to the sum of all the moments tending to turn the lever anticlockwise. Figure 57 shows a see-saw, the weight of the man 70 kg balances the weight of a young boy 40 kg and an older boy 60 kg.

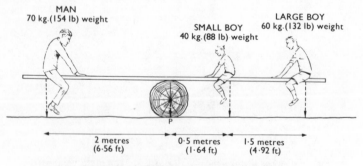

Fig. 57. The moment exerted by the weight of the man equals the moments exerted by the weight of the boys.

Remember 1 kg equals approximately 10 newtons (9·81).

Taking moments about the pivot *P*, the moment exerted by the weight of the man $= 700 \times 2$ Nm.

$= 1400$ Nm or 1·4 kilonewton metres.

The moments exerted by the weights of the young boy and the older boy

$$= 400 \times 0·5 + 600 \times 2 \text{ Nm}$$
$$= 200 + 1200 = 1400 \text{ Nm or } 1·4 \text{ kNm.}$$

Mechanical Advantage and Velocity Ratio

In many mechanisms, of which the lever is an example, an effort is used to move a load. In Fig. 54a a force *F* is shown lifting a load *W* by means of a lever. Two factors of the mechanism are:

Mechanical advantage. The mechanical advantage is measured by dividing the load by the effort.

$$\text{i.e. mechanical advantage} = \frac{\text{the load moved}}{\text{the effort exerted}}$$

In Fig. 54a the mechanical advantage $= W/F$.
(N.B. The above statement ignores any friction in the mechanism.)

Velocity ratio. The velocity ratio is measured by dividing the velocity at which the effort moves by the velocity at which the load moves.

$$\text{i.e. velocity ratio} = \frac{\text{velocity of the effort}}{\text{velocity of the load}}$$

If *A* is the distance moved by the effort and *B* is the distance moved by the load in any time *t*, the velocity of the effort will be A/t and the velocity of the load will be B/t.
In Fig. 54b the velocity ratio $= A/t$ divided by $B/t = A/B$.

Gear and Belt Drives

A gear wheel has a number of teeth cut around the circumference. If two such gear wheels have their teeth meshed together, when one of them rotates the other must also rotate, but in the opposite direction. Also, the speed at which they rotate will be in direct proportion to each other, see Fig. 58a. Supposing that the gear wheel *A* which has 16 teeth, is attached to a clutch shaft and drives in a clockwise direction, then the gear wheel *B* which also has 16 teeth will rotate anticlockwise. Because each of these wheels has the same number of teeth they will rotate at the same speed. This is because the same number of teeth on each wheel must pass the point where they mesh with each other, in the same time.

Supposing the drive gear *A* has 16 teeth and the driven gear *B* 32 teeth, Fig. 58b, then when 16 teeth of the drive gear pass the point where they mesh, 16 teeth of the driven gear will also pass. This means that whilst the drive gear has made a complete revolution, the driven gear has only made half a revolution. In order that the driven gear can make one revolution, the drive gear must make two revolutions. The wheels would have a speed ratio of 2:1.

Whenever a small gear wheel drives a large one, the large one will rotate at a slower speed. Whilst if a large gear wheel drives a small gear wheel, the small one will rotate faster.

If it is necessary to cause two gears to rotate in the same direction it can be done by fitting an idler gear between the two, see Fig. 58c.

The fitting of the idler gear will make no difference to the speed ratio of the two outside gears but will merely change the direction of rotation. The action of gears running together is similar to a series of levers.

A formula connecting the speeds of the two gears in mesh is as follows:

$$\frac{\text{speed of driving gear}}{\text{speed of driven gear}} = \frac{\text{number of teeth on driven gear}}{\text{number of teeth on driving gear}}$$

In Fig. 58b

$$\frac{\text{speed of } A}{\text{speed of } B} = \frac{\text{number of teeth on } B}{\text{number of teeth on } A}$$

It will be seen that the positions of A and B are reversed on the two sides respectively of the equation.

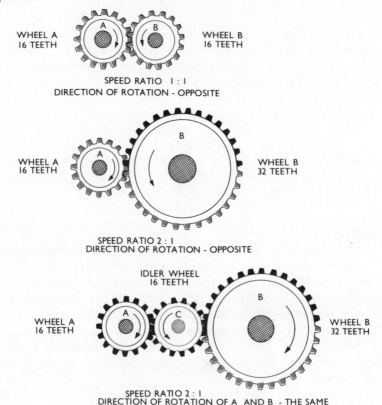

Figs. 58a, b and c. The principles of gearing.

Figure 59 shows a pulley G, of diameter g, driving a pulley N, of diameter n. If pulley G turns through one revolution, the belt will move distance πg; it will thus turn the edge of pulley N through a distance πg. Now since the distance round pulley N is πn, this pulley will turn through $\pi g / \pi n$ revolutions, which equals g/n revolutions, i.e. speed of the driven pulley = speed of the driving pulley $\times$ g/n which can be rewritten thus:

$$\frac{\text{speed of driving pulley}}{\text{speed of driven pulley}} = \frac{\text{diameter of driven pulley}}{\text{diameter of driving pulley}}$$

The relationship is exactly the same as for a gear drive: the only difference in the two forms of drive is that two gears in mesh will revolve in opposite directions whereas two pulleys connected by a belt will revolve in the same direction, unless the belt is crossed. *Note*: $\pi = 3 \cdot 142$ or $22/7$.

The term "speed" is not strictly correct; revolutions per minute should be used. However, it is felt that the word speed may, in the first instance, make this subject easier to understand.

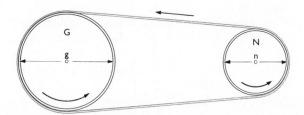

Fig. 59. The direction of rotation of the pulleys is the same; crossing the belt will cause them to rotate in opposite directions.

Examples

1. A gear with 15 teeth drives a gear with 75 teeth. If the speed of the driving gear is 55 r.p.m. what is the speed of the driven gear?

$$\frac{\text{Speed of the driving gear}}{\text{Speed of the driven gear}} = \frac{\text{number of teeth on the driven gear}}{\text{number of teeth on the driving gear}}$$

i.e. $$\frac{55}{\text{speed of the driven gear}} = \frac{75}{15}.$$

Therefore speed of the driven gear $= \dfrac{55 \times 15}{75}$ rpm

$$= 11 \text{ rpm}$$

2. An electric motor, fitted with a 100 mm (4 in. approx.) dial pulley and turning at 1400 rpm drives a hammer mill fitted with a 160 mm (6½ in. approx.) dia. pulley. At what speed will the hammer mill turn?

$$\frac{\text{Speed of the driving pulley}}{\text{Speed of the driven pulley}} = \frac{\text{diameter of the driven pulley}}{\text{diameter of the driving pulley}}$$

i.e. $$\frac{1400}{\text{speed of the driven pulley}} = \frac{160}{100}\left(\frac{6\cdot5\text{ in.}}{4\text{ in.}}\right)\text{approx.}$$

Therefore speed of the driven pulley $= 1400 \times \dfrac{10}{16}$ rpm

i.e. the hammer mill turns at 875 rpm.

Peripheral Velocity

The peripheral velocity of a pulley, or any other circular component, is the speed at which the rim is moving. This is calculated by imagining a point on the rim and working out how far this would travel in a given time. Shown in Fig. 59 is a pulley whose diameter is g units; suppose that it turns n revolutions per minute. In one revolution it will move a distance πg units and therefore it will move a distance πgn units when the pulley revolves n times. The peripheral speed is then equal to a distance πgn per minute. If slip is ignored, the speed of any belt driven by the pulley will be equal to the peripheral speed of the pulley.

Peripheral speed is often important in relation to farm machinery. For example, the peripheral speed of the threshing cylinder in a combine harvester should normally be about 1800 m/min (6000 ft/min approx.). In other words, the speed at which the rasp bars rub the crop to remove the corn from the straw is about 1800 m/min. (6000 ft/min approx.). It is this rubbing speed, rather than the revolutions per minute the cylinder makes, which is important. Some typical peripheral speeds of farm machinery and equipment are given below.

Grinding wheel peripheral speed 1800 m/min (6000 ft/min approx.)

Forage harvest peripheral speed 4000 m/min (13,000 ft/min approx.)

Circular saw peripheral speed 3400 m/min 1,000 ft/min approx.)

Examples

1. The threshing cylinder of a combine harvester, whose diameter is 530 mm (21 in. approx.), is driven at 1090 rpm. What is its peripheral speed?

Peripheral speed $= \pi dn$

$$= \frac{22}{7} \times \frac{530}{1000} \times \frac{1090}{60}\text{ m/s}$$

$$= 32\cdot26\text{ m/s}$$

or

Peripheral speed $= \pi dn$

$$= \frac{22}{7} \times \frac{21}{12} \times 1090 \text{ ft/min}$$

$$= 5995 \text{ ft/min.}$$

2. A 700 mm (28 in. approx.) diameter circular saw is to be driven from a tractor belt pulley, whose diameter is 300 mm (12 in. approx.) and which turns at 1000 rpm. What size pulley should be fitted to the spindle of the saw?

Peripheral speed of saw $= \pi dn$ (where d is the diameter of the saw in metres and n is the revolutions per minute made by the saw spindle).

$$3400 = \frac{22}{7} \times \frac{700}{1000} \times n$$

$$n = \frac{7}{22} \times \frac{3400}{1} \times \frac{1000}{700}$$

$$= \frac{7 \times 34 \times 1000}{154}$$

$$= 1545 \text{ rpm}$$

But $\qquad \dfrac{\text{rpm of driving pulley}}{\text{rpm of driven pulley}} = \dfrac{\text{diameter of driven pulley}}{\text{diameter of driving pulley}}$

$$\frac{1000}{1545} = \frac{\text{diameter of driven pulley}}{300 \text{ mm}}$$

$$\text{diameter of driven pulley} = \frac{1000 \times 300}{1545}$$

$$= 194 \text{ mm.}$$

A 194 mm dia. pulley should be fitted to the saw spindle,

or

Peripheral speed of saw $= \pi dn$ (where d is the diameter of the saw in feet and n is the revolutions per minute made by the saw spindle).

$$11,000 = \frac{22}{7} \times \frac{28}{12} \times n,$$

$$n = \frac{11,000 \times 7 \times 12}{22 \times 28} \text{ r.p.m.}$$

$$= 1500 \text{ r.p.m.}$$

But $$\frac{\text{rpm of driving pulley}}{\text{rpm of driven pulley}} = \frac{\text{diameter of driven pulley}}{\text{diameter of driving pulley}}$$

$$\frac{1000}{1500} = \frac{\text{diameter of driven pulley}}{12}.$$

$$\text{Diameter of driven pulley} = \frac{1000 \times 12 \text{ in.}}{1500}$$

$$= 8 \text{ in.}$$

An 8-in. pulley should be fitted to the saw spindle.

Force of Gravity

The earth exerts a force called the force of gravity on all matter, and thus if an article of any sort which has been suspended above the ground is released, it will fall towards the earth: The gravitational force acting vertically down on a body is known as its weight. This force acts, of course, on all parts of the article but it can be considered as acting vertically at a single point which is called the *centre of gravity* of the article.

Centre of Gravity and Stability

The position of the centre of gravity, often shortened to c.g., of an article is important to the machinery operator, particularly to the tractor driver because it governs the stability of the article. The c.g. of a rectangular box is at its centre: but when more articles or more complicated shapes are considered the position of the c.g. may be difficult to calculate. The farm machinery operator need not know how to do this calculation but he should know how the position of the c.g. affects stability.

Figure 60 shows a rectangular article which has been slightly tilted. If we consider the whole of the weight of the article to be acting at C, it will be seen that this weight produces a moment about P (the line about which the body has been pivoted) and that the article is tending to return to its original position. If, however, the article is tilted still further there will come a time when the c.g. is directly over P and the article should

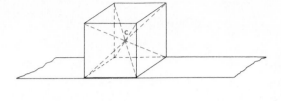

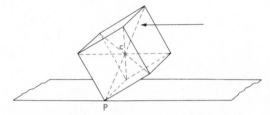

Fig. 60. The centre of gravity of a rectangular box.

balance. A slight push to the right will cause the article to return to its original position; but a slight push to the left will bring the c.g. outside the pivot point *P*; the weight of the body will now produce a moment about *P* which is tending to overturn the article and as there is no stabilizing force, the article will overturn.

These principles can be applied to the case of a trailer. Figure 61 shows the rear view of an unloaded two-wheel trailer; the c.g. is just above the axle and the trailer would have to be driven on a very steep slope indeed before the c.g. was outside the wheel track, the line about which the body is being pivoted, and the trailer would overturn. When the trailer is loaded, however, the c.g. is well above the trailer floor and the higher and greater the load the more the c.g. will rise. It will be seen in Fig. 61 that in this case it will need a less severe slope to cause the c.g. to come outside the wheel track and for the trailer to overturn.

In the case of four-wheel trailers similar principles apply but one important fact must also be considered. When travelling in a straight line along a slope, the trailer will tend to pivot about the front and rear wheels which are on the lower side of the slope: now if the tractor is turned to the left the front bogie will turn. This brings the front wheels towards the centre of the trailer and then the pivot point moves nearer the centre of the trailer and thus nearer to a vertical line through the c.g. so there is less stability. For this reason it sometimes happens that a four-wheel trailer may be quite stable when driven straight across sloping ground but it may overturn if the tractor driver turns either up or down hill. This may be difficult to understand until it is realized that when the front bogie of a four-wheel trailer is at right angles to its straight ahead position, the trailer becomes in effect a tricycle.

There is a further matter to be considered. Any moving body has momentum and if a trailer is driven fast round a corner a centrifugal force is produced, which acting at the c.g. causes an overturning moment. The higher the load, the greater will be the distance from the c.g. to the pivot point and thus the greater the overturning moment.

Fig. 61. The position of the centre of gravity on loaded and unloaded
trailers. When the centre of gravity acts through a line outside
the wheelbase the trailer will overturn.

Below are some points which should be borne in mind when using a trailer.

1. The higher the load, the higher the c.g. and the greater the tendency to overturn on a slope. Therefore the load should be as low as possible. Making a wider load will enable the same amount of material to be carried; but the c.g. will be lower.
2. The wider the wheel track, the greater will be the stabilizing moment exerted by the c.g. In hilly areas it is advisable to increase the wheel track; this can often be done by reversing the trailer wheels which are often dished.
3. The faster the cornering speed and the higher the c.g., the greater the overturning moment due to centrifugal force.
4. Particular care is needed when turning with a four-wheel trailer, both on a slope and also on level ground, because of the change of the position of the line about which the trailer will pivot. It is interesting to note that four-wheel trailers with rigid front axles do not suffer from this disadvantage; although this type of trailer has a more limited steering lock than a front bogie type, it is safer particularly on hilly land.

Stability of Tractors and Machines

The same principles apply to the tractor. The wheel track should be widened as far as practicable for use in hilly country and cornering speeds should be low.

Particular dangers arise when a fore-end loader is fitted because when the fork or bucket is filled and lifted, the c.g. of the whole outfit is raised. In difficult conditions it is advisable that the loader be lifted no higher than is absolutely necessary.

It should also be remembered that the c.g. of a combine harvester is normally raised when the grain tank is full, thus increasing the tendency to overturn. On some modern machines particular care has been taken to site these tanks as low as is practicable.

Density

The density of a substance is its mass (kg) divided by its volume (m³).

For example, if a block of silage has a mass of 200 kg and occupies a space of 1 m × 0·6 × 0·5 m then its volume is:

Volume = 1 × 0·6 × 0·5
= 0·3 m³.

∴ The density of silage would be

$$\frac{\text{mass}}{\text{volume}} = \frac{200}{0·3} = 666 \text{ kg/m}^3.$$

Specific Gravity (sp. gr.)

The specific gravity of a substance is its weight in relation to that of an equal volume of water. The specific gravity of water is considered to be 1 and the specific gravity of other substances are quoted as multiples and fractions of this. For example, the sp. gr. of diesel oil is about 0·85 or 85/100 of the weight of an equal volume of water. The sp. gr. of cast iron is 7·2, so its weight will be 7·2 times that of an equal volume of water.

CHAPTER 10

Electricity

ELECTRICITY is the flow of electrons (negative charges) in a conductor, from one atom to another. This flow of electricity can, in some ways, be compared with the flow of water.

Pressure (measured in volts) is the force causing the electricity to flow, and as with water, the greater the pressure, the faster is the flow. Electrical pressure can be provided by:

A generator—this is provided by the rotation of a magnetic field about a coil or winding. Typical examples of this are in the dynamo on a tractor or the generator at a power station.

A battery—two suitable dissimilar materials, usually lead and lead peroxide are situated in a chemical. An electrical pressure is created by the two materials which, if connected together, will result in the flow of a current.

Current (measured in amperes) will flow through a conductor, but we have to remember that for anything to move a resistance has to be overcome. Dependent on the degree of resistance to electrical flow we can classify materials into two groups:

Conductors—These are materials which offer little resistance and consequently will allow a current to flow through them. Gold, copper, steel, carbon and water solutions are examples of good conductors.

Insulators—If a material resists the flow of electricity and therefore will not conduct a current, it is referred to as an insulator. Wood, rubber, porcelain, glass, mica and air being the most commonly employed.

Resistance (measured in ohms) is present in all materials, though some more than others. With electricity, resistance shows itself as heat and this is power wastage, therefore for the conveyance of electricity we use wire with a low resistance. High resistance wire is used for such appliances as heaters when the heat produced is both desired and useful.

Ohm's Law

The relationship between Pressure, Current and Resistance is given to us by "Ohm's Law" which states:

The electrical current flowing in a circuit varies DIRECTLY with the electrical pressure and INVERSELY with the resistance. To put this in the form of an equation

$$\frac{\text{volts}}{\text{ohms}} = \text{amperes } or \text{ volts} = \text{amperes} \times \text{ohms.}$$

Example

The pressure in an electrical circuit is 12 volts. The resistance offered by the conductor is 3 ohms. What is the rate of flow?

Answer

$$\frac{\text{volts}}{\text{ohms}} = \text{amperes} = \frac{12 \text{ volts}}{3 \text{ ohms}} = 4 \text{ amperes}.$$

Circuits

For a current to flow, a circuit is essential. This means that there must be a complete line of conduction from the generator or battery, through the appliance and back to the source of pressure. On most vehicles it is the frame or chassis which provides the return to the battery, but in mains circuits two separate wires are used, live (delivery) and neutral (return).

Fuses

Protection of a circuit is essential for both the installation and the operator. Overloading of circuits can be caused by fitting too many appliances, by the use of multiple adaptors, using appliances with too great a current consumption, or using faulty appliances. All these instances can cause excessive current to flow, and, if this were permitted, overheating and possibly fire could result.

To limit current to a safe level fuses are installed in the circuit. These consist of a thin wire of a known current rating, 3, 5, 13 and 30 amp being the most common sizes. If this stated current rating is exceeded due to overloading as mentioned above, the fuse wire will melt or "blow" thus breaking the circuit and stopping the flow of electricity.

Fuses can be of the capsule type, as used in plugs or vehicle fuse boxes, or of the terminal type as used for heavier duty work, see Fig. 62.

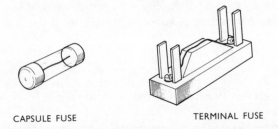

CAPSULE FUSE   TERMINAL FUSE

Fig. 62. Two types of fuses in electrical circuits.

When replacing fuses always remember:

(a) Switch off source of power at the mains or unplug the appliance.
(b) Replace the "blown" fuse with one of the same rating.
(c) If a fuse of the correct rating is continually "blowing" call in a competent electrician to ascertain the cause and put right the fault.

Some circuits incorporate mechanical fuses or cut-outs. These are switches which automatically break the circuit if the current becomes excessive. They can be reset at the flick of a switch once the fault has been rectified.

Power

Electrical power is measured in WATTS (after the Scottish engineer James Watt, 1736-1819). In practice the watt is often found to be too small a unit to be convenient therefore the kilowatt (kW) is often used.

1 kilowatt = 1000 W.

The watt may in some circumstances be referred to as a JOULE where:

Work done in Joules = Force (newtons) × distance (metres)

The approximate consumption of common appliances are:

Lights 40-200 W.
Heaters 750 W.
Grain-dryer air heaters 25 kW.

It is convenient at this point to mention that 746 watts = 1 horsepower.

Mains Electricity

Mains electricity is almost always 240-V, 50-cycle alternating current.

Single Phase

A single alternating current passes through the appliance; the voltage applied rises to +330 V then decreases to nothing. The voltage drops still further to −330 V, i.e. the pressure is in the opposite direction and it finally rises to nothing. This completes the cycle which takes place 50 times per sec. and is equivalent for power supply purposes to a supply of 240 V.

Three Phase

Three separate single-phase currents which are out of step, i.e. maximum and minimum voltages occurring at different times, are fed to the appliance. This produces a total voltage of approximately 415 V. Because of the staggered phasing, three-phase electricity offers many advantages for use in heavily loaded systems; large electric motors are nearly always of the three-phase type. However, for small motors and normal domestic supplies, single-phase systems are adequate.

Figure 63 illustrates the single- and three-phase alternating cycles.

The heat which is caused by the resistance to the flow of an electrical current is proportional to the square of the current. This means that if the current is doubled, the heat produced and thus the power loss is 4 times greater. Conversely, if the current is halved, the heat produced and thus power loss is reduced to a quarter of its original

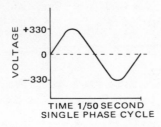 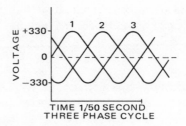

Fig. 63. Diagrammatic representation of single-phase alternating current and three-phase alternating current.

value. For this reason electricity which is to be transmitted over a long distance is usually converted to a very low current and incidentally to a very high voltage, and as the power (watts) is the product of pressure and current (volts × amps) there is very little power loss because the current (amps) is kept to a minimum. The conversion of voltage and current is carried out by a transformer.

The Transformer

Transformers are used to increase or decrease the electrical voltage from one circuit to another circuit belonging to a specific unit. When a current is passed through a conductor a magnetic "field" or region of electric strain is produced around the conductor. If a second conductor is placed in this field any changes in the field or any change in the position of the second conductor will induce an electric current in the second conductor.

This is known as electromagnetic induction and it is on this principle that electric motors, ignition systems and transformers operate.

In the transformer there are two coils, one called the primary winding and the other the secondary winding. An alternating current is passed through the primary winding and this produces a varying field which induces a current into the secondary winding. If the secondary winding has more turns than the primary winding the voltage is increased and the current proportionately decreased. If the secondary winding has less turns than the primary winding the voltage is decreased but the current is increased. Ignoring losses due to resistance there is no power loss during the transforming of electricity.

$$\frac{\text{Input voltage}}{\text{Output voltage}} = \frac{\text{no. of turns on p.w.}}{\text{no. of turns on s.w.}} = \frac{\text{output amperage}}{\text{input amperage}}$$

At the power station the current is passed in to a transformer where voltage is increased from approximately 240 V up to as much as 132,000 V, this reduces the current and minimizes loss due to heat as the power is supplied through the national grid system. On reaching the user the electricity is passed through a series of transformers where the reverse process takes place. Voltage is reduced (for safety reasons) and the current is increased. It would be impractical to use a very high voltage for electrical apparatus on the farm because of the degree of insulation which would be necesary.

Colour Coding

Even at 240 V electricity is dangerous if it is misused. A heavy current flowing through the body can cause severe burns and also electrocution of the operator. For this reason most electrical appliances are earthed. This means that the casing of the appliance is connected through a third wire to the socket and thence to the ground, then should any breakdown of insulation occur the current can "leak" to earth via the wire rather than via the operator. This leakage of current is sufficient to cause the fuse to blow which will indicate to the operator that the appliance is faulty and a thorough check should be carried out by a competent electrician.

We can therefore see that the installation of an appliance necessitates three wires, a feed or "live" wire, a return or "neutral" wire, and an earth wire.

It is essential that these wires are connected in the correct positions and to facilitate this they are coloured differently so that one can easily determine which wire is which. These colours are Brown, Blue and Green/Yellow striped.

The BROWN wire is used for the live part of the circuit and any switches should always be incorporated in this wire.

The BLUE wire is used for the neutral part of the circuit, but remember that this wire is also "live".

The GREEN/YELLOW striped wire is used for the earth line.

Figure 64 shows the correct method of connecting these wires to a modern three-pin fused plug.

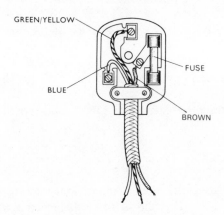

Fig. 64. Wiring connections of a three-point plug.

On older installations and appliances different coloured wiring may be found. These are still significant and are as follows:

    Red   live wire
    Black neutral wire
    Green earth wire.

Earthing

Although the correct earthing procedure may have been followed when three-pin plugs have been connected, safe working depends on there being a good circuit from the socket outlet to the earth. This circuit is often taken through metal tubing (or conduit), which protects the wires, to a plate sunk into the ground. This circuit is not always effective. The joints in the conduit may become rusted and sometimes the metal plate itself does not make good contact with the soil. The result of this would be that the appliance would become live if a fault occurred and the operator would receive a shock. It is essential that electrical systems be inspected regularly by a competent electrician in order that the earth continuity (efficiency of the earth circuit) may be checked.

Costing of Electricity

Electrical power is measured by multiplying together the voltage which is applied to a piece of equipment and the current which it passes.

Volts × amps = watts.

A power of 1000 W or 1 kW used for 1 hour is called a kilowatt hour or Board of Trade unit (usually shortened to a unit). The cost of these units varies considerably according to the amount consumed, the area in which used and possibly the time of day that it is used. Thus it is impossible to quote an accurate cost of electricity but an average figure is about 6 p per unit.

The following examples show how to calculate running costs.

(a) One 100-W light bulb used for 1 hour would cost 100/1000 kW × 1 hour.
= 1/10 unit.

(b) One 2000-W heater used for 1 hour would cost 2000/1000 kW × 1 hour = 2 units.

(c) One 18-kW grain drier heater used for 1 hour would cost 18 kW × 1 hour
= 18 units.

(d) One 1-h.p. motor used for 1 hour would cost 746/1000 kW × 1 hour = 0·746 unit.

(N.B. For practical purposes it is often assumed that a motor uses not 746 watts per h.p. but 1 unit per h.p. per hr. This allows for mechanical and other inefficiencies.)

Summary

Pressure is measured in volts.

Rate of flow is measured in amperes.

Resistance is measured in ohms. The resistance of a conductor is determined thus:

$$\text{Resistance} = \frac{\text{voltage applied to the conductor}}{\text{amperes which the conductor will pass}} = \text{ohms.}$$

Power is measured in watts. 1000 watts = 1 kilowatt.

Quantity is measured in units, 1000 watts being used for 1 hour is equivalent to 1 unit.

Electricity and the Tractor

Virtually all self-propelled implements have a battery. They need it to operate lights, electric controls, and more important for automatic starting. On petrol-engined vehicles the battery also has to provide the power for the coil-ignition system. This battery stores electrical energy in a chemical form and when required the energy can be released. The current supplied is d.c. (direct current).

Note. A.C. (alternating current) cannot be stored, it must be created as it is needed.

In order for a current to flow we must have a circuit. Figure 65 illustrates a simple electrical circuit. The terminals of a battery are joined together by a piece of wire through which electricity will flow. If a bulb is put in the circuit, it will light up and energy will be used up in doing this.

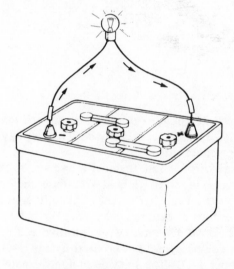

Fig. 65. A simple electrical circuit.

If the light was left on for a long period it would eventually use up all the electrical energy in the battery. The battery would then be in a discharged state, and before the battery could be used again it would have to be recharged.

Lead-acid Battery. Construction

The battery consists of a number of cells, each cell having an approximate electrical pressure of 2 V. The cell is made up of POSITIVE and NEGATIVE plates which hold the chemical materials in flat grids. Negative plates are made of lead (Pb) which is grey in colour and the positive plates contain lead peroxide (PbO_2) which is chocolate brown in colour.

A cell is made by welding a number of similar plates together on a terminal post, then plate groups of different polarity are interlaced so the plates alternate negative, positive and so on. Negative plate groups have normally one more plate than the positive group so as to keep the negative plates exposed on both sides (see Fig. 66). Each individual plate is then kept apart from its neighbour by porous separators. The whole assembly is contained in a battery case with a vent cap on top to allow the battery to be checked for electrolyte level and permit gases to escape.

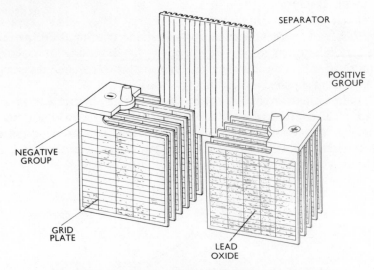

Fig. 66. Sets of plates interleaved together form a battery cell.

We now have a 2-V cell, the number of plates in that cell not affecting the voltage but determining the duty of the cells. A heavy-duty battery having more positive and negative plates than a light-duty battery.

To increase the voltage we must link together a number of cells. Six-volt batteries contain three cells connected in series, while 12-V batteries have six cells in series. This is how batteries come to us, a number of cells linked in series and should we want higher voltages, combinations of batteries are used. The battery works by a chemical reaction between the chemical materials of the positive and negative plates and the electrolyte which is dilute sulphuric acid. The electrolyte in a fully charged battery has a specific gravity of about 1·27. The solution is approximately 36% sulphuric acid and 64% water. Whenever it is necessary to mix such a solution, the acid must always be poured into the water, never the reverse.

When a battery is discharging, the current is produced by a chemical reaction which causes oxygen in the positive plates to combine with the hydrogen in the electrolyte to produce water (H_2O). At the same time lead in the positive plate combines with the sulphate in the electrolyte to give lead sulphate; from the negative plate lead combines with sulphate of the electrolyte to form lead sulphate. Thus, lead sulphate is formed at both plates when the battery discharges and the sulphuric acid is weakened by the production of water.

Charging of the battery which is done by the generator of the vehicle is basically a reversal of the discharge procedure.

Checking the Electrolyte Level

This should be carried out weekly as part of a maintenance schedule. The correct level is normally to the bottom of the filler neck, so that the plates are just covered. If the level is low, it should be replenished using distilled water. Acid should never be added to a battery unless spillage has occurred as this affects output voltage. The state of charge can be tested using an instrument known as an hydrometer.

Battery Capacity

The capacity of a battery is its ability to produce an electric current over a definite period of time. Because current is measured in amperes and time in hours, the battery's capacity is stated in ampere hours. This means that if a battery is said to be of 100 ampere hour rating, it should deliver 10 A over a period of 10 hours, or it could deliver 5 A over a 20-hour period.

The battery is an important part of the tractor ignition and starting equipment; in fact, the system will not operate without it. Many starting and ignition troubles can be traced back to battery neglect or misuse, therefore proper attention is essential.

The Ignition Coil

A coil is used in the ignition system of a spark ignition engine to increase the voltage supplied by the battery. It is necessary to increase the voltage in order that a spark may take place across the points of the spark plug situated in the combustion chamber. Battery voltage may be 6 V or 12 V but it is necessary to provide between 10,000 V and 12,000 V. This is because the spark must jump across the points against the high pressure within the cylinder.

The coil is in fact a small transformer consisting of a primary winding, a secondary winding and a soft-iron core. This soft-iron core forms the centre of the coil and the secondary winding of fine wire is coiled round it (see Fig. 28). This coil may have as many as 20,000 turns on it, the end of which is connected to the high-tension terminal. The primary winding is made up of about 400 coils of heavier wire wrapped around the secondary winding. The two ends of the winding are attached to the two terminals on the cap of the coil. The s.w. terminal being connected to the ignition switch wire and the c.b. terminal being connected to the wire leading to the contact breaker points.

Both the soft-iron core and the windings are shrouded with laminate and the unit is sealed into a metal container filled with either oil or an insulating material.

Operation of the Coil

When the contact breaker points are closed a circuit is set up allowing current to flow from the battery, through the primary winding to the contact breaker points, and then to earth. This establishes a magnetic field in the soft-iron core, and when the points open the flow of electricity stops; the magnetic field collapses, and in doing so a high voltage is induced into the secondary winding. This voltage could be as high as 20,000 V but it will only go as high as needed to cause the current to jump the spark gap.

The Dynamo

The function of the dynamo is to:
(a) supply the electricity needed by the vehicle,
(b) keep the battery in a charged condition.

The latter it can only do when the demands from electrical accessories (i.e. lights, heater, etc.) do not exceed the dynamo's output capacity.

The dynamo consists of a coil of wire, known as the armature, which is rotated between magnetic poles. As the armature rotates, voltage is generated and this is carried to the commutator. In order for current to flow a circuit must be set up and this consists of brushes which rub against the commutator and pass the electricity to the wires of the electrical circuit. Figure 67 shows an arrangement for a very simple dynamo.

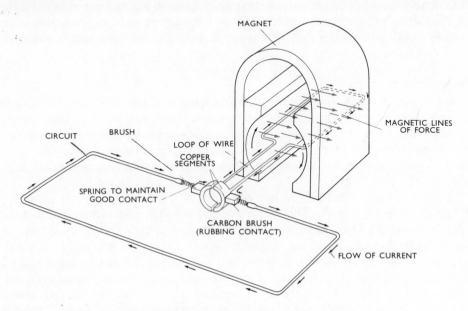

Fig. 67. An arrangement for a very simple dynamo.

The Armature

This is made up on a steel shaft on to which iron discs are pressed to form a core. These discs have slots around their circumference and it is into these slots that the coils of wire are held.

The Magnets

The types used are known as electromagnets and these consist of soft-iron blocks called pole pieces which have coils of wire wound around them. The coils of wire are called field coils and by passing a current through these wires we can control the strength of the magnets.

The Commutator

This is part of the armature and is made up of a number of copper segments, insulated from each other, to which the coils of wire are attached.

The Brushes

These are the rubbing contacts which are made of carbon and they collect the current which is produced and pass it on to the circuits.

The components are all housed in a casing, a "V" pulley is attached to the end of the armature shaft and the drive is taken via a "V" belt from the crankshaft pulley of the engine. The tension of this belt is adjustable by varying the position of the dynamo, facilitated by the three mounting bolts of the unit. Correct belt tension is achieved when the belt can be deflected approximately 1 in. under normal thumb pressure on the longest belt surface.

Under-tightening may result in belt slip which will reduce the output of the dynamo and could lead to deterioration of battery state.

Over-tightening will result in excessive wear of the armature shaft bearings and of the belt itself.

Operation of the Dynamo

Referring to Fig. 67 we see the basic units of the dynamo, the armature and the magnets. If the armature coil is rotated, the breaking of the magnetic field by the wire causes electricity to be produced.

The voltage generated at this stage cannot supply a current until a circuit is completed, and to do this the commutator and brushes are added. The brushes are spring loaded to maintain good contact with the commutator and current is now flowing in the circuit.

To increase the output of the dynamo, field windings are coiled around the magnets and it is by these windings that the output of the dynamo can be regulated.

We now have the completed dynamo which will supply electricity at the required voltage and amperage as is determined by the regulator in the charging circuit. The output of a dynamo is determined by:

(a) Number of coils on the armature.

(b) Speed of rotation.

(c) Strength of the magnets.

From these three factors it can be determined that by varying the current in the field windings we can alter the strength of the magnets and thus control output.

Figure 68 shows an exploded view of the main parts of a dynamo.

The Alternator

With modern machinery there is an increasing use of electrical equipment which places heavy demands on the battery and charging circuit of a machine, and it is this factor which has brought about the application of the alternator, which, although it is smaller and lighter than the dynamo, has the capacity to supply a higher current at low engine speeds.

The alternator employs the same principle as the dynamo in that when a magnetic field is broken by a coil of wire, electricity is produced. The dynamo, as we have seen, does this by moving the conductor or coil through a stationary field, whereas the alternator has a moving field (rotating magnet) within a stationary conductor.

This operation produces alternating current (a.c.) which is then rectified electrically to direct current (d.c.) for use in the vehicle electrical system.

Figure 69 shows the principle of operation of the alternator.

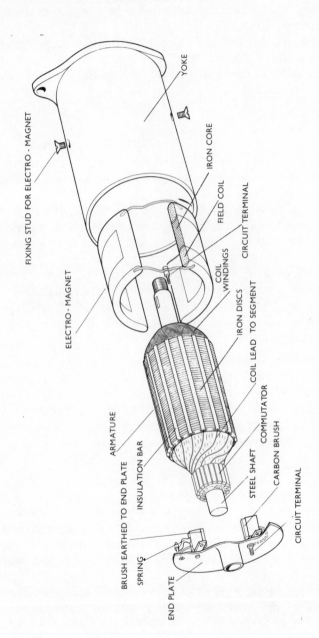

FIXING STUD FOR ELECTRO - MAGNET

YOKE

IRON CORE

FIELD COIL

CIRCUIT TERMINAL

COIL WINDINGS

ELECTRO - MAGNET

IRON DISCS

COIL LEAD TO SEGMENT

ARMATURE

COMMUTATOR

BRUSH EARTHED TO END PLATE

INSULATION BAR

STEEL SHAFT

CARBON BRUSH

SPRING

CIRCUIT TERMINAL

END PLATE

Fig. 68. The working parts of a dynamo.

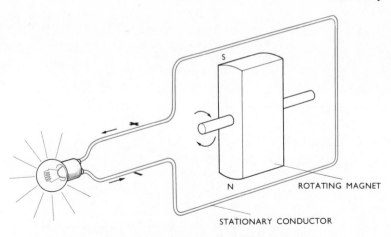

S

N   ROTATING MAGNET

STATIONARY CONDUCTOR

Fig. 69. The principle of operation of the alternator.

The Starter Motor

The principle of the starter motor is based on magnetism and it is constructed very much like a dynamo. It has field coils, armature and brushes, but its method of operation is the reverse of that of the dynamo. The dynamo is driven by the engine and it produces current which flows to the battery. When the starter motor is operated continually and the engine will not fire, the battery is very soon exhausted.

Passing a current through a wire conductor creates a magnetic field around the conductor and again, if the conductor is coiled around a soft-iron core, the magnetic field is intensified. This arrangement gives us an electromagnet as in the dynamo, but whereas in the dynamo the electromagnets are magnetized during manufacture by passing an electric current through them, in the starter motor current passes through them from the battery. This happens whenever the starter switch is operated and a strong magnetic field is created around them. Figure 70 shows four electromagnets such as would be used in a starter motor. They are the starter field coils and they are connected together in such a way that current flows through each one in turn. They are said to be connected in series. The manner of connecting them is such that they become alternate north and south poles of a magnet. Now if we can also pass the current through the armature, the armature coils will also produce a magnetic effect and will have a north and south polarity.

The armature in Fig. 70 is shown simply as two coils, the ends of which are attached to commutator segments. If when current passes through these coils they are given a north and south polarity, they will be caused to rotate because the north pole on a field coil will attract a south pole on an armature coil. With only one coil on an armature it would, of course, stop turning when the north and south poles came together. However, if there are many coils on the armature it is possible to get continuous rotation. This is because as one coil moves out of line of attraction another would move into it.

The armature is connected in series with the field coils by brushes which bear on the commutator, and as can be seen in Fig. 70 the current will pass through each armature coil in turn as the segments come in contact with the brushes. Two brushes only are shown, one of which is earthed to complete the circuit.

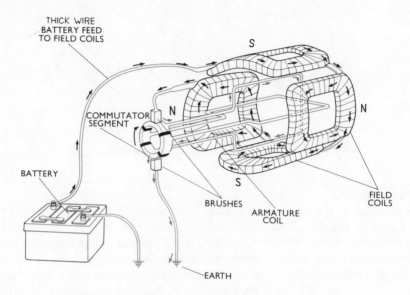

Fig. 70. The principle of operation of a starter motor.

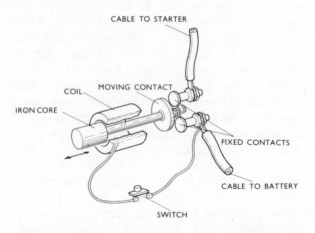

Fig. 71. The principle of operation of a solenoid switch.

Electrical Switches

Electrical switches are solenoids and are often employed in electrical circuits where motors using high currents are involved. To carry high currents we have to use thick, heavy cable which as well as being costly also absorbs energy in the form of heat losses, so in such cases the heavy duty cable is kept as short as possible. A solenoid switch is installed to open or close the circuit and this solenoid can be operated by a light duty switch as can be found on the control panel of a modern tractor.

Operation

The solenoid consists of a soft-iron core which is free to move, placed at one end of a coil of wire, both the coil and the core assume a similar polarity, therefore if a current is passed through the coil the core is drawn into the centre of the coil because adjacent poles are of different polarity.

It is on this principle that the electric switch operates for this movement of the core can be employed to close, at high speed, the two contacts of the heavy-duty cable (see Fig. 71). A spring is fitted in the switch to return the core to its initial position after the switch is released.

CHAPTER 11

Heat

MATTER is made up of atoms (in the case of elements) and molecules, i.e. combinations of atoms (in the case of compounds). These atoms move within a limited zone (and a very very small amount). As a body is heated, the vibration of these atoms increases; if the body is heated enough these vibrations give off rays of light which are dull red, bright red or even white if enough heat is applied.

Temperature and Quantity of Heat

To understand the difference between these two measures a simple experiment can be done. Light a match; the heat produced by this match is enough to burn one's finger if it is placed too near, yet it will not heat appreciably a kettle of water. But it may be possible to touch momentarily an electric hot plate which has been turned off and allowed to cool yet this plate can warm a pan of water. In the first case the match flame has a high temperature but there is only a small quantity of heat in the flame, in the second case the hot plate is at only a low temperature but contains quite an appreciable amount of heat.

The *temperature* of a body is a measure of the vibration of the atoms in the body. The *quantity of heat* in a body is a measure of its weight, the type of material of which it is made and its temperature.

Temperature Units

There are two common scales by which temperature are measured, *Centigrade* and *Fahrenheit*.

Centigrade: water freezes at 0° and boils at 100°.

Fahrenheit: water freezes at 32° and boils at 212°.

Conversion of Temperature Scales

To convert Centigrade C° to Fahrenheit F°, multiply C° × 9/5 and add 32,

e.g. $35°C = 35 \times \dfrac{9}{5} + 32°F$

$= 95°F.$

To convert Fahrenheit F° to Centigrade C°, subtract 32 from F° and multiply by 5/9, e.g. $104°F = (104 - 32) \times \frac{5}{9}°C$

$\qquad = 72 \times \frac{5}{9}°C$

$\qquad = 40°C.$

Quantity of Heat

There are two common units used to measure quantity of heat.

Calorie

The calorie is the quantity of heat needed to raise 1 gramme of water 1°C.

British Thermal Unit (B.t.u.)

The B.t.u. is the quantity of heat needed to raise 1 lb of water 1°F.

Specific Heat

If a given quantity of heat is applied to different substances they do not all achieve the same temperature rise. Specific heat is a measure of the quantity of heat needed to raise a given weight of a substance 1° Centigrade or Fahrenheit in relation to the quantity of heat needed to raise the same weight of water 1° Centigrade or Fahrenheit. For example, the specific heat of oil is 0·66: this means that to raise a given weight of oil 1° is 66/100 of the amount needed to raise the same weight of water 1°.

It will be seen then that the same quantity of heat will raise a given weight of oil to a higher temperature than it will raise in temperature the same quantity of water. Water is a better cooling agent than oil and although the latter does do some of the cooling in an engine, the water in the radiator and water jacket of a water-cooled engine absorbs most of the heat. Air is also a poor cooling agent and air-cooled engines only work efficiently when a large quantity of air is blown over the engine. This means that special attention should always be given to the exterior cleanliness of this type of power unit.

Effect of Heat

Expansion

When a body is heated it expands; allowance is made for this fact when engines and other machines are designed but the operator should be careful, particularly when using a new engine, that it is not overloaded and thus overheated. The expansion of the heated parts may cause the heavy loading between the moving parts which can lead to seizure. This is a result of the heavy loadings and heat breaking the oil film between the parts; when they come into contact small areas of the metals may melt and "weld" together. In severe cases the engine may "lock up" but in any case the bearing surfaces will be damaged.

Change of State

Matter can exist as solid, liquid or gas. In the first case the atoms vibrate through only a small range and the body retains its shape. If the body is heated, the range of vibration of the atoms increases until they can move anywhere in the substance, in fact the body liquefies. More heat causes the atoms to vibrate still more until individual atoms vibrate outside the substance and leave it altogether; the substance boils and turns into gas.

The most common substance in which these changes can be observed easily is water.

Latent Heat

When a body is heated the temperature rises until it reaches the temperature at which it changes its state. Now any extra heat that is used to effect this change of state is absorbed until all the material in the body has been changed from solid to liquid or liquid to gas; only then does the temperature rise again. The heat necessary to cause this change of state of a material is called its latent heat. This fact can be shown quite easily by measuring the temperature of boiling water. Although heat is continuously applied, it will be noted that the temperature of the water does not rise; the extra heat is being used to cause a change of state of the water from the liquid state to the gas state.

Transmission of Heat

Heat is transmitted by three means.

Conduction

If a poker is put into a fire not only the end in the fire gets hot but in time the handle also will become too hot to hold. The heat is being conducted along the poker because the vibration of the atoms of iron which are in the fire is gradually causing the atoms of iron in the handle to vibrate. When the handle is gripped in the hand the vibration of the atoms in it is transmitted to the atoms in one's hand.

Heat is transferred by conduction between any articles in contact which are at different temperatures.

Not all substances have the same heat-conducting power. Copper is very good; for this reason copper and brass (a copper alloy) are used for tractor radiators and aluminium and its alloys, which are also good conductors, are often used for the cylinder heads of air-cooled engines. Materials such as felt and glass fibre are poor conductors, i.e. insulators, and they are very useful for "lagging" hot water pipes and cisterns.

Convection

It has been noted that materials expand when they are heated. If a liquid is heated at the bottom those parts of it which are closest to the point of heat application expand most, they thus become less dense and rise; the cooler, more dense parts then fall to the bottom and they in turn become heated, expand, become less dense and rise. Thus a circulation is set up. Heat is transmitted then by convection through liquids and gases due to the heated parts becoming less dense and rising.

Radiation

When the atoms in a body vibrate, due to the heat contained in the body, they emit heat waves in the form of radiation. When these waves meet another body they make the atoms in that one vibrate also thus raising the temperature. If one stands close to a tractor which has been working hard, one can feel the heat without actually touching the engine.

In a tractor engine the heat is transmitted:

(a) By conduction from the cylinders and head through the metal walls to the water which transfers it by conduction to the radiator tubes and fins, and thence to the air drawn through the radiator by the fan.

(b) By convection due to the heated water rising and flowing through the top hose to the top of the radiator: cooler water flows through the bottom hose to replace this hot water; the circulation is usually assisted by an impeller.

(c) By radiation from all parts which are hotter than the surrounding air.

These processes can be made less efficient by scale or rust on the inside of the water jacket which lowers the conduction of heat; by dirt or hard water deposit in the radiator tubes which slows the convection currents, and by exterior dirt which lowers the radiation of heat from the exterior of the engine.

CHAPTER 12

Materials

THE materials of which tractors, farm implements and equipment are made can be grouped into two main classes, namely *metals* and *non-metals*.

Metals

Metals may be used in various ways.

1. As pure metals.
2. As substances containing non-metals, of which carbon is the most common.
3. As substances containing two or more metals; these are called *alloys*.

Metals are known as *ferrous* when they are composed largely of iron and *non-ferrous* when they do not contain iron.

Ferrous Metals

WROUGHT IRON

Wrought iron is almost pure iron. It is relatively weak and has a *tensile strength* of about 3150 kg/cm² (20 tons/in²); it is not used much in the construction of farm machinery. However, pieces of wrought iron can be joined easily by heating them white hot and hammering them together, a process known as forge or fire welding. For this reason, it is very suitable for making chain links and ornamental gates and fences.

MILD STEEL

Mild steel is a mixture of iron and about ½% (1 part in 200) of carbon. This material has a tensile strength of about 4725 kg/cm² (30 tons/in²) and it is thus stronger than wrought iron. Mild steel can be formed into round or rectangular bar, angle iron and sheet metal by rolling it, when red hot, between suitably shaped rollers. Much of the framework and sheet metal work of farm machines is made of mild steel.

CARBON STEEL

When the amount of carbon mixed with iron is greater than about ½%, the material is usually called carbon steel. The carbon content varies slightly in different types of carbon steel: and steels containing up to about 1% (1 part in 100) are in common use. Carbon steel is stronger than mild steel and has a tensile strength of up to about 6300 kg/cm² (40 tons/in²). Also carbon steel can be hardened and tempered (made softer and tougher) to give the material the properties it needs for particular uses.

Carbon steel, sometimes just called steel, is used to make the framework of cultivation machinery, hoe blades, the disc coulters on ploughs, forage harvester flails and other parts where strength is essential.

ALLOYS

Metals are sometimes used in the pure state but they are often found mixed with other metals in the form of alloys. When particular mixtures of metals are melted the atoms arrange themselves in definite patterns and the resulting alloys have different properties from the separate metals which make up the alloys. These metals may contain iron and other metals or they may be composed of two or more non-ferrous metals.

ALLOY STEELS

Metals such as nickel, chromium, tungsten and manganese can be added to steels to produce *alloy steels.* The quantities added are small, seldom more than a few parts in a hundred but these metals have a great effect on the steel produced. For instance, the addition of nickel gives a very tough steel which is used for axle shafts; while chromium is used to make stainless steel. Tungsten steel is used for good quality twist drills and manganese is added to make a steel which is resistant to soil wear.

HARDENING AND TEMPERING OF STEEL

Hardening. When steel is heated to red heat, there is a change in the arrangement of the iron and carbon particles and if the material is *quenched,* that is cooled suddenly (usually by plunging it into water), this new arrangement is maintained and the steel is found to be very hard. Unfortunately, it is also very brittle and must usually be *tempered* (made softer and tougher) before it can be used.

Tempering. When the hardened steel is warmed, the structure changes and if the material is quenched the new structure is "fixed" and the material will be found to be softer and tougher. The tempering temperature is much lower than that used for hardening the steel and it is seldom above 280°C. The higher the temperature used in the tempering process, the softer and tougher the material becomes.

Annealing. If hardened steel is heated to red heat and allowed to cool very slowly, it becomes soft. This process, which is called *annealing,* is often used in the manufacture and repair of farm machinery; so that parts made of steel can be drilled or turned on a lathe.

Case hardening. This process consists of adding carbon to the surface of a low carbon steel, thus making hardening possible. The object to be case hardened is packed in a special material and heated, when the carbon is absorbed into the surface. The object is later heated red hot and quenched, usually in oil, and a hard skin or "case" results. The thickness of this skin or "case" is normally only about 1·25 mm (0·050 in. approx.) or about 1/16 in. While the surface is very hard, the centre of the object does not become brittle as the composition has not been altered. This process is often used for hardening engine and gearbox components; but it is not suitable for parts which are in contact with the soil; because as soon as the "case" was worn away the rest of the part, being soft, would not last very long.

CAST IRON

Cast iron is the name given to a range of materials which contain iron and about 3-4% of carbon. When this mixture of iron and carbon is heated it becomes molten at about 1200°C and the molten metal can be poured into sand moulds, a process known as *casting*. The moulds are usually prepared by packing special sand and binding agents round a wooden pattern of the object to be made, the whole being contained in a metal box. The mould from which the pattern has been removed is then baked in an oven when the sand hardens: after cooling, the pattern is removed. The molten metal is poured into the resulting cavity and on cooling it sets in the required shape. In this way complicated objects can be made cheaply.

Cast iron can exist in two forms; they both contain the same constituents but the speed of cooling affects the structure produced. If the cooling is quick, *white cast iron* is produced, but if the cooling is slow, *grey cast iron* results.

WHITE CAST IRON

When cooled quickly from the molten state cast iron is hard and brittle and if it is broken the faces show a white crystalline structure. This process is often used to make parts which have to stand soil abrasion. A steel plate or *chill* is built into the sand mould and the molten cast iron lying against this plate cools quickly and thus takes on the hard form. When ploughshares are made, chills are used to produce hardened surfaces on those parts which tend to wear fastest.

GREY CAST IRON

When cast iron is cooled slowly from the molten state some of the carbon separates from the mixture. If the material is broken the faces show a grey colouring structure due to the presence of this carbon, which occurs in the form of graphite. Grey cast iron is softer and stronger than white cast iron and has a tensile strength of about 2205 kg/cm² (14 tons/in.²). Also, due to the slow rate of cooling there is less risk of distortion and cylinder blocks, gearbox and transmission casings of tractors and other articles of complicated shape, where adequate strength can be provided easily, are usually made of grey cast iron.

While grey cast iron is stronger than white cast iron it is weaker than steel and mild steel. Thus cast iron is not suitable for parts which have to withstand heavy forces unless these parts are very heavily made. However, cast iron may be heat-treated to produce castings which are much stronger than the original material.

MALLEABLE CAST IRON

When cast iron castings are heated for a long period in suitable packing materials, carbon is removed and a product much stronger than the original results. This material is called *malleable cast iron* and it is used to make many of the smaller lighter castings on farm implements. Complicated parts can be made cheaply of cast iron and although they would not be strong enough if left in this form, the alteration from cast iron to malleable cast iron which takes place during the heat treatment results in a component which is perfectly satisfactory.

Non-ferrous Metals

ALUMINIUM

Aluminium is used either pure, as sheet aluminium, or mixed with silicon to make castings. The aluminium sheet is weather resistant and some grain storage bins are made of this material. Aluminium silicon alloys are easy to cast and this is one of the reasons why this substance is used for tractor engine pistons and parts such as injection pump casings. Some tractor sumps are also made of aluminium silicon alloys.

BRASS

Brass is an alloy of copper and zinc. It is a good bearing material and is often used for the brass bushes which are found on nearly all farm implements. Water pump parts are also sometimes made of brass, due to its resistance to rusting.

BRONZE

Bronze is an alloy of copper and tin; like brass it is a good bearing material. Sometimes phosphorus is added to improve its properties, the substance is then called *phosphor bronze.*

COPPER

Copper is generally used alloyed with zinc or tin to give brass or bronze; although copper water pipes are often found in farm buildings. Also due to the fact that it conducts efficiently both heat and an electric current, copper is used for making tractor radiators and for the windings in electrical apparatus such as starters and dynamos.

LEAD

Lead is used for making the plates in tractor batteries. Some water pipes are also made of lead; but the material is much dearer than it used to be and iron or copper pipes are now more commonly fitted. Lead is alloyed with tin to make *solder,* which is used for the construction and repair of electrical apparatus and plumbing installations.

TIN

Tin, which is not usually used in the pure state now, owing to its very high cost, is a constituent of solder.

ZINC

Zinc is used as a coating for the mild steel plates known as *galvanized iron.* The clean steel plates are dipped in a bath of molten zinc which coats the surfaces. The zinc is resistant to corrosion by the weather and these sheets are used for the roofs of farm buildings and to make buckets, feeding troughs and other pieces of farm equipment. Fabricated articles should be constructed before being galvanized, as if they are constructed with material which has already been galvanized the coating is likely to be damaged during manufacture.

Zinc can be alloyed with magnesium, aluminium and copper to make alloys which melt at a very low temperature. These alloys can be cast, in steel moulds called dies, to a high degree of accuracy. The material is often called *zinc base die cast* and it is used to make carburettor parts and coil ignition distributor bodies. Complicated components can be made cheaply and easily but the alloy is not very strong and great care is necessary when handling this material.

Non-metals

Wood

Various types of wood are used in the construction of farm implements and equipment. Four common home-grown timbers, their properties and uses, are listed below.

Ash: a light strong resilient wood particularly suitable for fork and hoe handles.

Elm: a heavy tough wood durable in water often used for trailer bottoms.

Larch: a light resinous soft wood with some degree of resistance to decay used for fencing stakes and for some parts of huts for livestock.

Oak: a heavy hard wood used for gate posts and for those parts of construction work where strength is essential.

The cost of paint prevents its wide use for the protection of farm woodwork. Lead-based paints should never be used on equipment for stock as they tend to lick it and the intake of even a small amount of paint may cause death. Manufacturers can supply various types of paint which can be applied with safety to woodwork with which stock will come in contact.

Creosote is often used as a preservative; this has little effect, however, unless the timber is creosoted under pressure or heated in a creosote bath. But creosote is often applied to buildings and sheds for housing stock because of its disinfectant action.

Rubber

Rubber is used for making tractor tyres, driving belts (particularly those of the V type) and hose pipes. It is unaffected by damp but perishes in strong sunlight and it is attacked by fuel and lubricating oils. Rubber tyres can become permanently deformed if the vehicle or implement is left standing for a long time with the tyres under-inflated. To obtain the maximum life from tyres, the pressures should be kept at the correct figure: and if the vehicle or implement is to be stored it should be jacked up and left standing on blocks, the tyres being covered with sacks. Any fuel or lubricating oil which is spilt on a tyre or other rubber parts should be wiped off at once.

Plastics

The plastics most commonly found on the farm are those of the polythene type which are used for hose pipes, buckets and various water installation parts such as ball valve floats. These plastics are completely resistant to cold water, but they become soft and will be distorted by hot water.

Fibreglass, which is made of glass fibres bonded together by special adhesives, is often used for.belt and chain guards. The material is not affected by the weather but cracks may develop if the fixing bolts are not kept tight. If guards made of fibreglass are damaged they can be repaired easily by building up the broken areas with fresh glass fibre and adhesive. Sometimes it may be necessary to cut away the damaged parts; but full instructions are always given with the repair kits, which are readily available.

CHAPTER 13

Workshop Tools

Hacksaws

Hacksaw blades are available in lengths of 203 mm (8 in. approx.), 254 mm (10 in.), 304 mm (12 in. approx.); the shorter blades are more rigid and thus less likely to be broken but they do not cut as quickly as the longer ones. For beginners a 254-mm (10 in.) blade is quite satisfactory if care is taken; but when some skill has been acquired a 308 mm (12 in. approx.) blade is more satisfactory.

Blades can be bought whose number of teeth per 2·54 cm (1 in.) vary between 18 and 32. The coarser ones are suitable for soft materials in normal sizes, while for harder materials and for thin pieces of metal the finer blades are better. The table given below gives the recommended blades for different purposes.

Selection of Hacksaw Blades

Materials to be cut	Teeth per 2·54 cm (1 in.)
Small solid sections of soft materials such as mild steel, aluminium, brass and copper	
Large solid sections of hard materials such as alloy steel, heat-treated steel and stainless steel	18
Heavy angles	
Cast iron	
Small solid sections of hard materials such as alloy steel, heat-treated steel and stainless steel	24
Sections between 3 mm (1/8 in. approx.) and 6 mm (¼ in. approx.) thick, such as heavy tubing and sheets and medium angles	
Sections less than 3 mm (1/8 in. approx.) thick, such as thin tubing and sheets also light angles	32

The normal blades are made of a low tungsten steel; while they are comparatively cheap they do not last as long as the high tungsten blades sometimes known as "high-speed" steel. This does not mean that one should saw at a faster rate, but that the material from which they are made is a "high-speed" steel. These blades are much more expensive and they tend to be rather brittle, but in experienced hands they are much cheaper in the long run.

Files

Files may be either cut or double cut. On the former the teeth lie diagonally across the file blade but on the latter two rows of teeth cross diagonally (see Fig. 72).

Files with various numbers of teeth per inch are available ranging from very coarse to very fine, but for farm work two grades are in common use, namely bastard and second cut. The former is a medium coarse texture while the latter is rather finer.

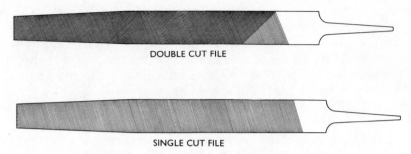

Fig. 72. Two types of workshop files.

Uses

Single cut files, which are normally supplied for farm work in the second cut grade are used for filing hard materials such as mower knife sections, wood saws, hand hoes, etc. They can be bought in flat or triangular shapes.

Double cut files are used for general work, the bastard grade being employed for removing metal quickly while if a finer finish is required a second cut file is used.

Care of Files

The teeth on a file are quite brittle and become easily blunted if the file is thrown into the assortment of tools generally found in a tractor toolbox. Ideally they should be wrapped in a rag or paper, not touching each other, if their life is to be prolonged. Damp also damages files as the crests of the teeth tend to rust away thus making them blunt. Oil should not be used on files as it tends to make the metal particles clog between the teeth. If this clogging occurs, files may be cleaned by gently brushing along the teeth with a wire brush or file card (a series of wire bristles fixed to a cloth backing).

Hammers

Figure 73 shows a selection of types of hammers; of these the ball pene is the best for the farm machinery operator. The ball end being used for riveting and spreading. Several points should be borne in mind when using a hammer. Firstly, the handle should be clean, dry and free from oil or grease; this enables the operator to strike

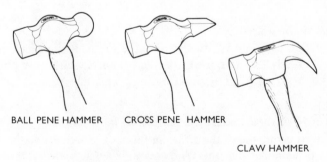

Fig. 73. A selection of hammers.

more truly and thus avoid expensive and possibly painful "misses". Secondly, it should be remembered that the hammer face is hardened and it may chip if used on hard materials. Also it will tend to deform anything which is struck, and when doing jobs such as knocking out a shaft, a piece of soft material should be used as a drift between the hammer and the object which is being struck. Thirdly, a hammer with a loose head should not be used. If the head has come loose, it can be driven farther on to the shaft and the wedge knocked down to secure it. Proper wedges should always be used and not a nail.

Special hammers are available, with heads made of copper or lead, for hammering articles which can be easily damaged, and hammers with rubber heads for tyre fitting.

Screw Threads

There are unfortunately many different types of screw thread in common use on farm machinery and the operator should know that these various types exist in order that he can make sure that correct replacements are ordered and fitted.

Generally, the number of threads on a bolt depends on its diameter. The greater the diameter the fewer threads per centimetre or inch. Bolt diameters are normally an exact number of millimetres, but the British and American systems are normally an exact number of 1/16ths of an inch, e.g. 1/4, 5/16, 3/8, 7/16, 1/2, etc., in the British and American systems but the Metric system uses bolts whose diameters are in an exact number of millimetres.

Metric

There are a number of metric threads in use, all of which are similar. The Systéme Internationale (S.I.) is a metric thread having the crest of the thread flat and the root rounded. The International Metric Fine Thread is similar but the pitch is smaller. French Metric threads are similar to the International system and sizes between 6 mm and 90 mm are the same as the International in thread form.

The German Metric Thread is also similar to the International.

Whitworth

This is a coarse thread particularly suitable for threads cut in weak materials such as cast iron or aluminium alloys: it is often used for general work on implements both for nuts and bolts and for adjusters.

British Standard Fine (B.S.F.)

This is a finer thread than Whitworth and due to this it can exert greater tightening force. B.S.F. threads are often used to assemble parts which have either to be tightly held together or are under a lot of vibration.

Comparison of Whitworth and British Standard Fine Threads

The table given below shows the different number of threads per inch for a selection of bolts. It will be noted that the Whitworth always has less threads per inch (for a given size of bolt) than does the B.S.F. bolt.

Diameter of bolt	Threads per inch Whitworth	Threads per inch B.S.F.
1/4 in.	20	26
3/8 in.	16	20
1/2 in.	12	16
3/4 in.	10	12
1 in.	8	10

British Association (B.A.)

This is a fine thread, only used on the small nuts and bolts normally used for electrical apparatus and instruments. Small bolts with a B.A. thread are occasionally used for fixing sheet metal parts. B.A. bolts are identified by numbers; a table is given below showing the diameters of various B.A. bolts. The numbers of threads per millimetre are also shown.

B.A. number	Diameter of bolt (millimetres)	Threads per millimetre
0	6·0	1·0
2	4·7	1·2
4	3·6	1·5
6	2·8	1·9

There is no need for the student to try and remember the particulars such as threads per inch, relating to different types of the thread; but he should know the general position.

Materials

Nuts and bolts are made of various materials.

Mild Steel

Black or bright mild steel is widely used for making bolts for general work where no great loads are involved.

High Tensile Steel

This material is used for bolts which have to undergo great loads; they can be distinguished by letters embossed on the bolt heads giving sometimes the maker's name and sometimes a figure such as T45 which shows that the bolt material has a breaking strain of 7087·24 kg/cm² (45 tons/in²). Care should be taken that any replacement bolts are of the correct type of material; machines can be easily damaged if this is not done. A particular case in point is the shear bolt fitted as a safety device to most balers.

Brass

The small B.A. bolts fitted to electrical gear are often made of brass to resist corrosion.

Galvanized Steel

Mild steel bolts, such as those used for assembling roofing gutters, are often galvanized to cut down corrosion.

Spanners

Many types of spanner are available but the open-ended, ring, socket and box types are the more commonly used for the maintenance and repair of agricultural machinery.

Open-ended

This type of spanner is relatively cheap and perfectly suitable for most jobs. It has two drawbacks, however; firstly, it is not always possible to use one of this type of spanner in a confined space, and secondly, the jaws are inclined to spring open under heavy loads. The correct method of using an open-ended spanner to exert a large force is shown in Fig. 74a, tightening a nut, and in Fig. 74b, loosening a nut. Note how the spanner has been turned over for the second operation. When used in this way, the tendency of the jaws to "spring" has been cut to a minimum.

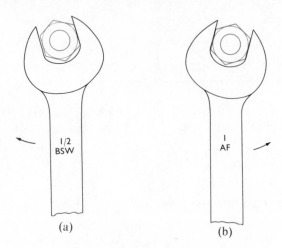

(a)   (b)

Fig. 74a. Tightening a nut.
Fig. 74b. Loosening a nut. See p.125 for spanner sizes.

Ring

Ring spanners are dearer than the open-ended type but they have two advantages; firstly, they are very suitable for use in confined spaces as the nut can be turned through 1/12th of a turn at a time, and secondly, the ring cannot open. Consequently, providing that the nut is not damaged, great force can be exerted without any risk of the spanner slipping.

Socket

Socket spanners are similar to a ring spanner except that they are of limited depth so that they cannot be used on a nut if a great length of bolt is protruding. However, the great variety of handles to which the socket can be fixed makes this type of spanner very adaptable. A selection of handles is shown in Fig. 75.

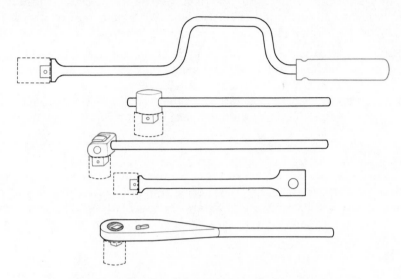

Fig. 75. A selection of handles for use with socket spanners.

Box

Box spanners are particularly suitable for nuts fitted in confined spaces, especially when the bolt protrudes any great distance through; they are also very good for removing and replacing sparking plugs which are often deeply socketed into an engine. Like ring and socket spanners they cannot "spring" so that large forces can be exerted.

Spanner Sizes

Spanners are made in different sizes to suit the range of nuts and bolts fitted to farm machines. Unfortunately, various systems are in common use and not all bolts of the same diameter are fitted with the same sizes of bolt head and nut.

The *American system* uses nuts which are an exact fraction of an inch across the flats. Consequently, a spanner to fit an American type nut will be marked with a figure such as "¾ in. A.F.". This means that the width between the jaws of the spanner is ¾ in. and that the spanner will fit a nut which is ¾ in. across the flats.

The *Metric system* is similar: a spanner marked 14 mm has a gap of 14 mm between the jaws and such a spanner will fit a nut which is 14 mm across the flats.

The *British system* is more difficult to follow. The width across the flats of the nut is 1·5D + 0·06 in. where D is the diameter of the bolt. This means that a spanner marked ¾ in. will fit the nut on a bolt which is of ¾ in. diameter. It is thus a much larger spanner than one which is marked "¾ in. A.F.". A further complication arises. Some years ago it was the custom to make nuts on bolts threaded with a Whitworth thread 1·5D + 0·16 in. across the flats while the nuts on bolts threaded with a British Standard Fine thread were 1·5D + 0·06 in. across the flats. This meant that the same size of spanner would fit either the nut on a Whitworth bolt or the nut on a British Standard Fine threaded bolt which was one size larger. Consequently, a spanner was marked thus

"3/8 in. B.S.W." "7/16 in. B.S.F.", which meant that the particular spanner would fit the nut on a 3/8-in. diameter bolt with a Whitworth thread and also the nut on a 7/16-in. bolt with a British Standard Fine thread. However, the Whitworth nut sizes were later brought into line with the British Standard Fine nuts; nevertheless, many spanners still retain this dual marking.

To sumarize. If a spanner is marked "¼ in. B.S.W." (or just "W") and "5/16 in. B.S.F.", it means that this spanner will fit the nut on a bolt whose diameter is 5/16 in. whether the bolt is threaded Whitworth or British Standard Fine.

Fortunately, a lot of modern machinery uses the American system in which the size quoted on the spanner relates to the width between the spanner jaws.

"Allen Keys"

Figure 76 shows a selection of "Allen keys" for use with screws which have recessed hexagon heads. They are the only satisfactory method of dealing with screws of this type; attempts to work with the tang of a file are never satisfactory. The keys themselves are specially heat-treated and are very cheap so that attempts to improvise are a waste of time.

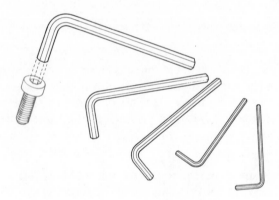

Fig. 76. A selection of "Allen keys".

Cold Chisels

Two types of cold chisel are shown in Fig. 77. The flat type is used for general work and the cross-cut type is used for cleaning out key ways; it is also very useful for general work; due to its shape it provides a very strong narrow chisel for use in confined spaces.

Note in Fig. 77 the correct shape of the cutting edge of a chisel. The end should have a gradual taper ending in a more abrupt taper at the actual cutting edge.

Cold chisels are usually sharpened on an emery wheel and care should be taken to keep the edge cool by frequently dipping it in water. Overheating will turn the cutting edge blue, which is a sign that the chisel has been softened. Some cold chisels are available which can be sharpened with a file; but they are rather hard and filing should be done slowly with a steady pressure using preferably a single cut second cut file.

Wood chisels are more delicate than cold chisels and should be sharpened on an oil stone and not on an emery wheel.

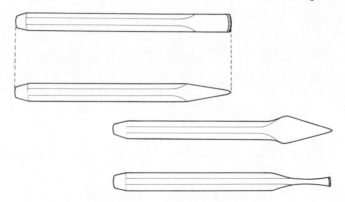

Fig. 77. Flat and cross-cut chisels for general work.

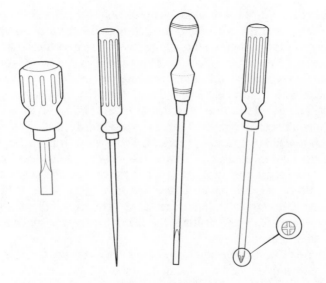

Fig. 78. A selection of screwdrivers.

Screwdrivers

Several types of screwdrivers are shown in Fig. 78. Note particularly the dumpy type, also the cross-ended type for use with recessed head screws.

The blade of a flat screwdriver should have a gradual taper but the end should be flat, the side faces almost parallel; the thickness should be slightly less than the slot width in the screw head. Sharpening is done by holding the screwdriver against the emery wheel as shown in Fig. 79a. A refinement is to finish off by grinding as shown in Fig. 79b. This ensures that the outside edges of the blade fit into the bottom of the screw slot and improves the grip. When the blade becomes too stumpy, the tapered faces can be ground on an emery wheel, taking care to keep the blade cool by dipping it frequently in water. The original taper angle should be maintained.

Cross-ended screwdrivers cannot be easily sharpened but due to their very positive method of gripping they rarely become unserviceable.

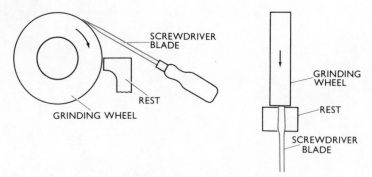

Figs. 79a and b. Reshaping a screwdriver blade.

Drills

The engineer's twist drill is commonly used for making holes in the course of repairs to farm machinery and the construction of fittings for farm buildings. These drills are made either of high carbon steel or "high speed" steel (which is a steel with a high tungsten content). The latter type, although dearer, is the better for drilling metal. Drills can have two types of shank either parallel or taper.

Figures 80a, b and c show a parallel shank drill and the machines for which it is suitable. It will be noted that the hand drill is of the two-speed type which is suitable for most farm work: the electric drill can have two speeds also. The ability to select a suitable drilling speed is important; excessive speed when using a large drill results in the drill becoming overheated and soft and too slow a speed when using a small drill makes the job take longer than necessary. When fitting a parallel shank drill into a chuck great care should be taken in making sure that it is tight. When a key-type chuck is used it is advisable to tighten the chuck in at least two of the three holes provided. If the drill is allowed to slip when drilling, the shank becomes scored and it is then more difficult to keep it tight.

After much use drills become blunt; they can be resharpened but this operation needs a lot of skill and unless one knows how to do it the drill may be made worse and possibly be damaged permanently.

Locking Devices (Fig. 81)

A great many devices are used to prevent nuts coming undone. Notes on a selection of these are given below.

Spring Washers

Spring washers prevent nuts coming undone because the sharp edges on the ends of the split part dig into the nut and the object (being held) respectively. As they dig in they resist the tendency of the nut to loosen. It should be noted that spring washers "wear out". When the edges become rounded or the springs collapse, the washers lose their efficiency. "Star" locking washers work in the same way as spring washers; but due to the fact that they have a number of twisted, sharp-edged tabs they are more effective.

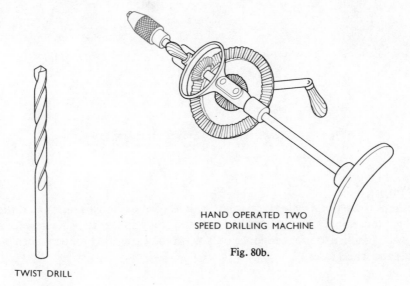

HAND OPERATED TWO
SPEED DRILLING MACHINE

Fig. 80b.

TWIST DRILL

Fig. 80a.

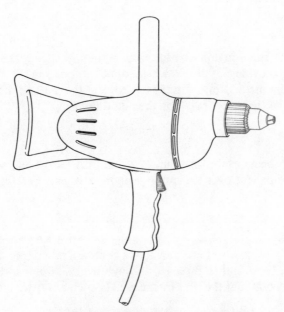

POWER OPERATED HAND DRILLING MACHINE

Fig. 80c.

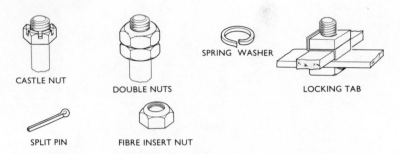

Fig. 81. A selection of locking devices used to prevent nuts coming
undone.

Split Pins

Split pins, sometimes called cotter pins, can be used with castle nuts and drilled bolts
to prevent loosening. They should be replaced if they appear to be damaged: in fact it
is the usual practice to use them once only when they are fitted to items such as engine
big ends and wheel axles.

Locking Nuts

Sometimes a second nut can be screwed on to the main nut on a bolt to stop the
assembly from becoming loose.

Self-locking Nuts

Some nuts incorporate a nylon or fibre ring in the top edge. The bolt-threads cut
into this ring and provide a "drag" which opposes loosening. Another type has
minute slots or a thin metal capping in the top edge of the nut and the whole is shaped
so that the top part of the nut "drags" on the bolt thread.

Locking Tabs

A piece of drilled sheet metal is fitted under the nut. After the nut has been tightened,
one edge of the tab is bent over the nut and another edge is bent over the object on
which the nut bears.

Castle Nuts and Wire

When a series of nuts have to be retained, it is common practice to drill the bolt ends
and use castle nuts. Wire is then threaded through the bolt holes and the ends twisted
together. It is important that the wire is threaded in such a way that it tends to tighten
all the nuts.

The above is only a selection of devices and the student will meet others. It should
be remembered that any locking devices have been put there for a purpose and great
care should be taken in replacing them correctly; considerable damage and even an
accident may result if this is not done.

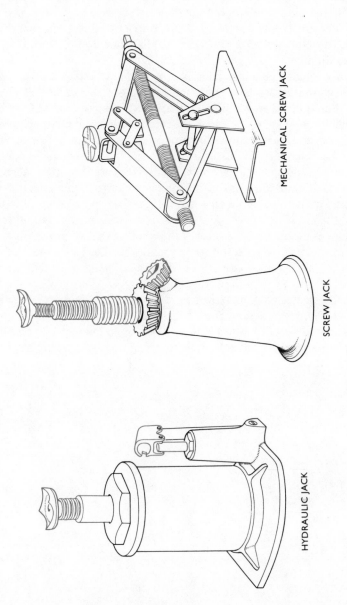

MECHANICAL SCREW JACK

SCREW JACK

HYDRAULIC JACK

Fig. 82. A selection of jacks.

Jacks and Jacking

Figure 82 shows a selection of jacks. Screw jacks are relatively cheap but they tend to be hard work to operate due to the high frictional losses in the system. Hydraulic jacks either of the bottle type or the trolley type are much more convenient to operate, but they cost more than screw jacks of the same lifting capacity. Particular advantages of the trolley jack are its low height when in the down position, the fact that loads can often be wheeled about when on the jack and that they do not tend to topple over if the load moves slightly.

It is quite easy to damage a hydraulic jack by trying to pump it up above its proper height; and the manufacturer's operational instructions should be followed carefully.

Great care should be taken when jacking up machines, especially if there is any risk of the load falling on anyone. If any work is to be done under a machine, then when the machine has been raised it should be supported by wooden blocks of adequate size so that it may be held as rigidly as possible, and no work should be done under any machine which is supported only by jacks. Particular care should be taken when using hydraulic jacks that there is no risk of the lowering control being operated by mistake.

Extreme care should also be taken when lifting wheeled machines. When one part has been lifted the machine may roll and displace the jack: tractors and other implements should have blocks placed on both sides of the wheels to safeguard against this danger. Where the machine must be lifted on soft ground, then the jack should be placed on a wooden block or short plank to assist in preventing it being squeezed downwards into the soil.

It is very bad practice to gain extra height by putting a jack on a brick; the latter can easily tilt or crumble and allow the load to fall.

Screw jacks should be well lubricated in order to lessen the effort involved. Hydraulic jacks oil reservoirs may occasionally need topping up and only the correct grade of oil should be used.

CHAPTER 14

Ploughs

THE continuing improvement in the performance and control of tractor hydraulic systems has now led to the disappearance of trailed ploughs, even when associated with crawler tractors the provision of hydraulic lift arms has meant that ploughs now fall into two categories:

(a) *Mounted.* The plough is attached directly to the three-point linkage and although on some ploughs the depth may be controlled by a land wheel, it is far more common for the implement to be controlled by the hydraulic's "draught-control" service. Although the design of such a system only assures a constant load upon the tractor it is fair to assume that as long as the soil texture and therefore the draught force remain constant, so too will the ploughing depth.

(b) *Semi-mounted.* Increases in the number of plough bodies has meant that the tractor is incapable of lifting the implement, so in order to maintain control, the plough is attached only to the tractor's lower links which are responsible for lifting the front of the implement whilst the rear is catered for by a "castor" type wheel which runs on the unploughed land. To lift the implement out of work the tractor hydraulics are raised, lifting the front of the plough; a hydraulic ram then exerts a force on the rear wheel which causes the rear end of the implement to lift clear of the soil. Depth control of the rear of the plough may also be incorporated in a mechanical screw handle acting on the rear castor wheel.

The layout of a typical small fully mounted plough is illustrated in Fig. 83. The main components are:

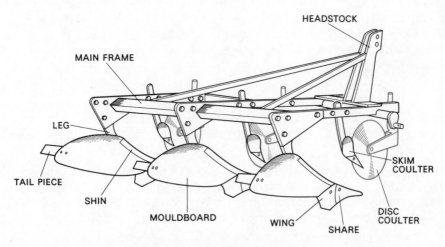

Fig. 83. A three-furrow mounted plough.

133

The Main Frame

The main frame is frequently made of tubular steel of either rectangular or round cross-section. Apart from providing an extremely strong backbone to which other components may be attached it also provides a simple assembly on to which additional bodies may be attached or removed, thus providing a plough which is ideally matched to the tractor power available. Plough "body" is the name given to the complete assembly responsible for turning a furrow and comprises of:

The Share

The share, or point as it is often referred to, is responsible for making the horizontal cut under the furrow slice and starts the lifting movement which is continued by the mouldboard. Shares may be made of cast iron or steel. The former being slightly cheaper and more suitable to soil types which are relatively stone free. The cast iron share is "chilled" on the underneath and landside edges, thus forming a white cast iron in these areas which is harder than the remainder of the share. The result is that the share wears away more rapidly on the top edge, giving a self-sharpening effect and also retaining its digging angle or "suck" for a longer period. Steel shares have the advantage of being less liable to damage from stones than cast shares and also they may be built up and "hard surfaced" when worn.

For use under extremely abrasive conditions a "bar-point" share can be used. This consists of a carbon steel bar which projects through from the rear of the body. As the point of the bar wears, thus reducing the plough's ability to penetrate, the steel bar can be moved forward to give the correct profile. Bar point assemblies are usually spring loaded to absorb shock when ploughing in hard and difficult conditions.

The Mouldboard

The mouldboard is responsible for the shattering and inversion of the furrow slice. The degree of both of these factors being determined by the type fitted:

(a) *General-purpose.* Associated with shallow ploughing, approximately 200 mm deep (8 in.), in grassland, clay soils or where there is little depth of soil. The body is long with a gentle curvature and a cross-sectional convex curve. The work it produces is unbroken and consequently requires considerable secondary cultivation to produce a seed-bed. This type is better used for winter ploughing when maximum furrow surface is exposed for weathering.

(b) *Semi-digger.* Associated with deep ploughing, approximately 250 mm deep (10 in.), for root crops or where the establishment of a seed-bed soon after ploughing is desirable. The body is shorter than the G.P. type, with a greater curvature and a concave cross-section. The work produced is more broken, requiring less secondary cultivation.

(c) *Digger.* Associated with very deep ploughing, approximately 300 mm deep (12 in.), for root crops or land reclamation work. The body is very short and twisted giving maximum shatter and inversion and is very concave in cross-section. The work produced is very broken, with individual furrow slices being virtually indistinguishable.

(d) *Other types of mouldboard.* Although the above types of mouldboard as described are mainly in general use, other types are available from manufacturers and they may be designed to produce a particular effect in particular soils and conditions. Effects produced will depend on the length of mouldboard, degree of twist, concavity or convexity. Obviously many conditions intermediate to those described can be produced.

The Disc Coulter

The disc coulter makes a vertical cut in the soil, thus separating the furrow slice from the unploughed land. Incorrect setting of the disc can not only cause bad ploughing but considerably increase the "draught" of the plough. The correct setting should be such that the centre line of the disc should fall just behind the point of the plough. There should be a vertical clearance of approximately 40 mm (1·5 in.) and a sideways clearance of approximately 15 mm (5/8 in.) between the disc and the plough point. Under hard conditions, however, the disc may act as a wheel preventing penetration. To help overcome this situation the disc may be raised and moved rearways.

Where ploughing is to be carried out under conditions of loose surface trash (e.g. straw), a smooth conventional disc may push the material in front of itself. Under these circumstances wavy-edge or scalloped discs may be used. For ploughing under ideal conditions, "knife edge" plough shares are now being used. Here the conventional share and disc is replaced by an integral share with a vertical knife on the landside edge. This single component producing both the horizontal and vertical cut. Disc coulter setting faults:

1. *Ragged furrow wall:* the disc is set too close to the ploughing, resulting in the share tearing beyond the cut of the disc. On light soils this may result in the collapse of the furrow wall.
2. *Stepped furrow wall:* the disc is set too far away from the plough, resulting in a step where the soil tears to meet the cut of the share.

The Skim Coulter

The skim coulter is fitted to improve the burial of trash. It does this by removing a small furrow slice approximately 102 mm (4 in.) wide and 70 mm (3 in.) deep, from the landside corner of the main furrow slice. This is deposited into the bottom of the furrow trench and subsequently buried by the inversion of the furrow slice. The clearance between the side of the skim coulter and the disc should be as close as possible without causing obstruction and preventing rotation of the disc.

The Landside

The landside presses against the furrow wall absorbing the sideways thrust of the plough; the rearmost body is fitted with an extended landside, because at this point the sideways force is greatest, and to this landside is fitted a detachable "heel". This component will wear in time and can, therefore, be replaced.

The Frog

The frog may be either a casting or a pressing to which the share, mouldboard and landside are attached. The frog is bolted to the beam and on some ploughs provision may be made for altering the pitch, which causes the downwards "suction" of the plough. Excessive pitch, however, will result in premature wear, rough furrow bottoms and poor tractor steering.

Plough Attachment

Prior to the attachment of a plough it is necessary to ensure that the tractor-wheel widths match the working width of the implement, otherwise it will be impossible to obtain the correct front furrow width. Because of specific manufacturer's designs it is important to consult the maker's handbook as to wheel settings but the following table may be used as a guide.

Width of furrow (mm)	Rear tyre size (mm)	Wheel width setting (mm)	
240 (10 in.)	264 × 864 (11 × 36 in.)	1344	(56 in.)
	(12 × 38 in.)	1344	(56 in.)
288 (12 in.)	264 × 864 (11 × 36 in.)	1344	(56 in.)
	288 × 912 (12 × 38 in.)	1440	(60 in.)
	336 × 816 (14 × 34 in.)	1440	(60 in.)
336 (14 in.)	264 × 864 (11 × 36 in.)	1440	(60 in.)
	288 × 912 (12 × 38 in.)	1440	(60 in.)
	336 × 816 (14 × 34 in.)	1536	(64 in.)

Attachment of the implement should be carried out in the conventional manner for mounted machines which is:

1. Reverse the tractor, with the lower links raised just above the level of the cross-shaft, towards the plough until the left-hand lower link is correctly placed for attachment. Lower the linkage until the left-hand lower link is at the correct height.
2. Brake the tractor and dismount.
3. Attach left-hand lower link and secure with linchpin.
4. Using the levelling box fitted to the right-hand lower link and if necessary the cross-shaft adjusting screw, align the right-hand link, fit and secure with linchpin.
5. Fit top link between headstock and tractor, adjusting as necessary.
6. Remount the tractor and raise the implement using the tractor hydraulic system.
7. Excessive swing of the implement may cause it to foul the tractor wheels; this may be prevented by shortening the check chains evenly on each side of the tractor, ensuring that there is still sufficient free-play to permit the plough to move freely when in work.

Note. To prevent possible damage to tractor cabs care should be taken when lifting implements.

Plough Controls

1. *Front Furrow Width*

This is controlled by the twisting of the cross-shaft. This is achieved by the winding of a screw handle attached to the cross-shaft (see Fig. 84a) or on some ploughs it may

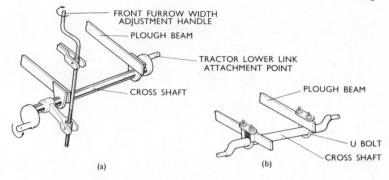

FRONT FURROW WIDTH ADJUSTMENT HANDLE

PLOUGH BEAM

TRACTOR LOWER LINK ATTACHMENT POINT

CROSS SHAFT

PLOUGH BEAM

U BOLT

CROSS SHAFT

(a)

(b)

Figs. 84a and b. Front furrow-width adjustment.

be controlled by an hydraulic ram worked from the tractor's hydraulic system. The effect of rotating the cross-shaft is to move its cranked ends causing the plough to swivel out of line with the tractor. However, as soon as the tractor and plough move forward, the plough will move across to run in a straight line, thus widening the front furrow width if the cross-shaft is twisted in a clockwise direction, or narrow the front furrow width if turned anti-clockwise.

2. *Pitch*

This is controlled by the length of the top link. As an approximate guide the overall pitch of the plough is correct when the "heel" of the rear landside is gently rubbing the bottom of the furrow. Shortening the top link will increase pitch; excessive adjustment will produce a rough furrow bottom and poor steerage; lengthening the top link will reduce pitch but again excessive adjustment will now prevent penetration. A typical top link is shown in Fig. 52 (e).

3. *Sideways Level*

This is achieved by adjusting the levelling box situated on the right-hand tractor lift arm. To compensate for the tractor's offside wheel running in the furrow bottom, the right-hand link arm is shortened until the plough is running parallel to the land.

4. *Plough Depth*

This may be controlled in a number of ways.

(a) WHEEL CONTROL

If the plough is fitted with a depth wheel the position of the wheel can be set by a screw handle, see Fig. 85. Raising the position of the depth wheel produces deeper ploughing and lowering the depth wheel produces shallower work, as the plough will continue to penetrate until the wheel rides on the soil surface.

(b) DRAUGHT CONTROL

(i) *Top link sensing.* When a plough is in work the draught force of the soil pushes against the plough bodies. Because the implement is attached to the tractor's lower

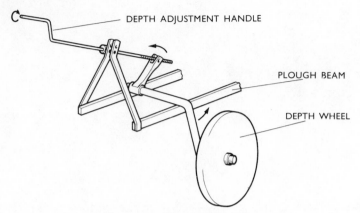

Fig. 85. Depth wheel adjustment.

links a pivot is provided resulting in the force on the plough bodies being transferred in the opposite direction down the top link. As the top link bracket is free to move, against the resistance of a spring, any increase in plough depth results in a related increase in top link pressure, this via linkages inside the tractor, will lead to the hydraulics raising the implement. Similarly should the implement ride shallow in the ground the force on the bodies becomes less. The pressure on the top link is reduced and the internal linkage permits oil to leak out of the ram cylinder, thus lowering the implement. Through this control, therefore, the draught force on the plough remains fairly constant.

(ii) *Lower link sensing.* The advent of larger mounted implements and the increasing use of semi-mounted implements has led manufacturers to adopt lower link sensing on large tractors. The lower link arms of the hydraulic linkage are attached to a common bar running across the width of the tractor. As a plough is pulled through the ground the force on the bar causes it to distort, the amount being related to the draught on the implement. This distortion can then be used to control the hydraulic system raising or lowering the implement to ensure the correct predetermined load. The principle of operation of a lower link sensing unit is shown in Fig. 48.

Reversible Ploughs

Basically the reversible plough consists of a conventional right-hand set of bodies mounted on the plough frame, with a corresponding number of left-hand bodies, mounted upside down, above them. By simply raising the plough out of work and turning the plough over, which can be done mechanically on smaller ploughs or hydraulically on larger models, the opposite set of bodies can be brought into work. A two-furrow reversible plough is shown in Fig. 86.

Adjustments

The setting of disc coulters, skim coulters, pitch (top link length) and depth (draught control) are identical to that of a right-hand plough. However, care must be taken to ensure both sets of bodies are evenly matched.

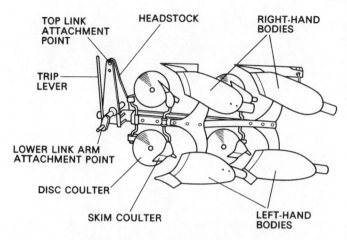

TOP LINK ATTACHMENT POINT

HEADSTOCK

RIGHT-HAND BODIES

TRIP LEVER

LOWER LINK ARM ATTACHMENT POINT

DISC COULTER

SKIM COULTER

LEFT-HAND BODIES

Fig. 86. A two-furrow reversible plough.

Sideways level and front furrow width are somewhat different and firstly it is important to remember that because both sets of tractor wheels will be running in the furrow bottom as the tractor ploughs in both directions the tractor must be symmetrical. Tyre pressures must be identical in each rear wheel, rear lift arm length should be checked and matched by the adjustment of the levelling box. Now the level of the plough is controlled by the adjustment of the stop on the turn-over mechanism; it is necessary to adjust this setting in both directions.

Front furrow width is determined by firstly ensuring that the plough is centrally mounted on the cross-shaft, this being positioned by the sliding clamps, then with the plough in work, to increase the cut of the right-hand front body, loosen the offside furrow width adjuster and tighten the off-side adjuster. Once the setting has been determined in one direction it will also be correct for the left-hand set of bodies.

Ploughing Methods

There are a few different methods of ploughing a field. These are:

(a) Systematic ploughing; where the field is ploughed in lands.
(b) Round-and-round ploughing; where the field may be ploughed starting from the outside and working to the centre, or starting from the centre and working to the outside.
(c) One-way ploughing; where the ploughing starts at one side of the field and is continued until the other side of the field is reached. A plough known as a reversible plough is used for this method of ploughing.

It is not within the scope of this book to discuss these ploughing methods in detail but a brief outline is given.

Systematic Ploughing

In this method of ploughing the type of plough used is that which turns the furrows in one direction only, that is, to the right. The plough may be mounted or trailed and single or multi-furrow ploughs may be used.

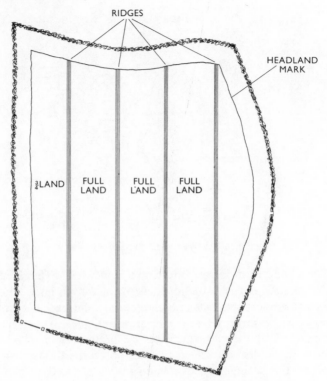

Fig. 87. A field marked out for ploughing in lands.

Because the plough turns the furrows in one direction only it is necessary to mark the field out in such a way that the ploughing is done in blocks or "lands". By doing this, the minimum amount of time is spent with the plough travelling out of work. Marking out of the field is commenced by ploughing a marking furrow around the field at a suitable distance from the field boundary. This distance varies according to the size of plough and tractor used, but it should be of sufficient width to allow easy turning of the equipment between the marking furrow and the boundary. There is nothing to be gained by having this mark (headland mark) too near to the boundary. The headland width may be between 6 metres and 12 metres (6.5 yd and 13 yd approx.) according to the size of equipment used. For example, a tractor using a two-furrow mounted plough can turn on a 6- or 7-metre (6.5-7.5 yd approx.) headland but a crawler tractor using a five- or six-furrow trailed plough will require about an 11-metre (12 yd approx.) headland width.

After deciding the width that the headland should be, sticks are set up at this distance at suitable intervals around the field. These then act as guides for the ploughman who steers his tractor in direct line with them and removes them as he comes to them on his way around the field. The marking furrow is cut with the rear mouldboard and it should be shallow, 8 cm (3 in. approx.) because its purpose is to serve as a guide mark where the ploughman lifts and lowers his plough when ploughing the lands.

The lands are marked out across the field and these should be a suitable width which is again largely determined by the size of tractor and plough used. Suitable land widths would be as follows:

Size of equipment	Land width (metres and yards)
Tractor and two-furrow plough	20-30 m (21-32 yd approx.)
Tractor and three-furrow plough	20-40 m (32-43 yd approx.)
Tractor and four-furrow plough	40-50 m (43-54 yd approx.)
Tractor and five-furrow plough	50-60 m (54-65 yd approx.)

In this case it does not matter whether the plough is trailed or mounted. The first land is usually made three-quarters the width of a full land. Figure 87 shows how the field would be divided up into lands.

The lands are divided by a ridge, sometimes called a rig, opening or split. Two types of ridges can be made: (a) arable ridge and (b) grassland ridge.

The arable ridge is better used when marking lands in fields that were previously cropped with an arable crop such as corn, potatoes, sugar-beet, etc. When this type of ridge is made, all the land beneath it is ploughed. The grassland ridge is used when marking lands in grass fields. Figures 88a and 88b show the different stages of work in ploughing these ridges and how the furrows lie when they are complete. It will be noted that in the grassland ridge a strip of land is unploughed.

It is generally better to set out all the necessary ridges in a field before the main ploughing is done. By doing this, the different operations in setting up a ridge can be carried out on each ridge in turn without altering the plough setting.

When all the ridges are made the lands are ploughed out starting with the three-quarter land. This is reduced to a quarter land width by ploughing between the ridge and the headland mark. Ploughing starts down the side of the ridge and down the headland side and is continued until the land is reduced to the quarter land width. The tractor and plough are turned to the left each time they come out or go into work. The operation is known as *casting*.

When this land has been reduced to the right width, ploughing then takes place around the ridge. The tractor and plough should now be turned to the right each time

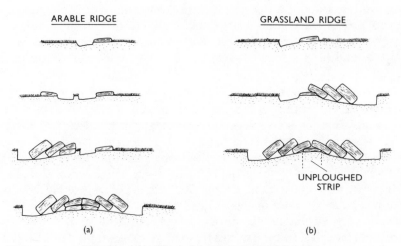

Figs. 88a and b. The various stages in ploughing ridges.

they come out or go into work. This operation is known as *gathering*. If ploughing is continued in this manner the three-quarter land width will eventually be completed and the first full land will be at the same time reduced to a three-quarter land width. The same procedure is then carried out with each land until they are all ploughed out. Generally, the last two lands have to be reduced in turn to a quarter land width when they are then finished together by ploughing first along one and then along the other.

Finishes (open furrows) have to be made in the last stages of ploughing out each quarter land.

The headland remains to be ploughed and this should be ploughed in the opposite direction to which it was the previous year. This is done to prevent the forming of a deep furrow or high ridge around the position of the headland mark as would happen if the headland was ploughed in the same direction each year.

Round-and-round Ploughing

This method of ploughing is also carried out with the right-hand plough and it is commonly practised where the conditions are suitable. The setting up of a lot of ridges is avoided, likewise a lot of finishes, but it is better carried out where (a) the fields are of good shape and (b) the fields are fairly level.

Fields that are of complicated shape or very undulating may be difficult to mark out.

Of the two ways in which this method of ploughing can be done, probably starting from the centre and working to the outside is the most satisfactory. In order to carry out this method it is first necessary to have marked out in the field centre a plot of land, say, 1 hectare (2.4 acres), of the same shape as the field itself. This can be done by ploughing a number of marking furrows around the field at different distances from the boundary. A useful method of doing this is as follows:

(a) A shallow headland mark is drawn in the normal way and at the normal distance from the boundary.
(b) With the aid of another man and a length of twine, a second mark is drawn by the plough. This is done by tying the twine to some point on the tractor and with the helper holding the other end and walking along the first marking furrow the tractor driver travels so that the twine is kept taut. Thus, the plough makes another furrow mark at an equal distance all round from the headland mark.
(c) More furrow marks are made in the same manner until the small plot in the field centre is marked out, which should be of similar shape to the field boundary, see Fig. 89.

The distance that these marking furrows are apart can be 30-40 metres (32-43 yd) approximately. The wider the distance the less accurate is the marking out likely to be.

When this marking out is completed, the centre plot is ploughed out by setting a ridge along its centre and ploughing around it. The plough is lifted or lowered each time it comes to the marking furrow. Having completed this, the rest of the field is ploughed round and round by travelling along each side of the centre plot and lifting the plough out of work and making a loop turn at each corner, see Fig. 89. As each marking furrow is reached the ploughing can be trued up.

This method of ploughing results in some of the land at the corners where turning takes place being ploughed twice. For this reason it is not generally carried out in grass fields because the turf is left exposed at the corners.

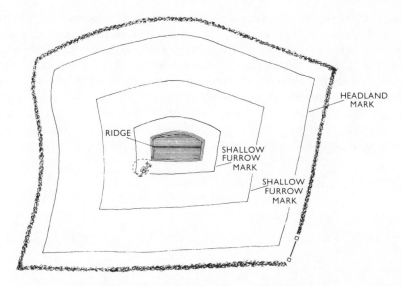

Fig. 89. A field marked out for "round-and-round" ploughing.

The headland is ploughed in the usual manner and should again be ploughed in an opposite direction each year.

To plough a field starting from the outside and working towards the centre, marking out is not normally done. The ploughman travels with his tractor and plough along the field boundary and turns the furrows towards the boundary. When he gets to the end of the field site, he must lift his plough out of work so that he can turn and move across to the next side before dropping the plough into work again. This must be done each time he completes a ploughing run along a side and the result is that diagonal strips are left across the field which must be ploughed after the bulk of the field is ploughed. These strips will be about 8 metres (18 yd approx.) wide and, although they are not difficult to plough, it will be found that before they are finished being ploughed a lot of the ploughed land will be paddled down by the tractor.

One-way Ploughing

This method of ploughing requires the use of a *reversible* plough, that is, a plough that is fitted with two sets of mouldboards, one set turning furrows to the right and the other to the left, only one set being used at any one time. One-way ploughing is a very satisfactory method and it has many advantages over the other methods. These are:

(a) Marking out of the field is simple and straightforward.
(b) No ridges or finishes required, thus fields are kept free from humps or depressions.
(c) Ploughing starts at one side of the field and continues by working back and forth across the field until the other side is reached.
(d) When fields that slope have to be ploughed the reversible plough will plough all furrows down hill if required.
(e) The time taken to plough a field is reduced because less time is spent travelling out of work.

The field is marked out with a headland mark only, this being of suitable width to allow easy turning of the tractor and plough, see Fig. 90. Alternatively, two sideland marks can be made on opposite sides of the field. Ploughing commences along the straightest side, the plough being lowered into work at the headland mark and lifted again at the headland mark on the opposite side of the field. The first furrows should be ploughed towards the boundary. When the opposite headland mark is reached and the plough lifted, a reverse turn is made, see Fig. 90. The plough is turned over and ploughing is continued down the same furrow with the mouldboards turning furrows in the opposite direction. This procedure continues until all the field except the headland is ploughed. The headland can then be ploughed in the normal manner.

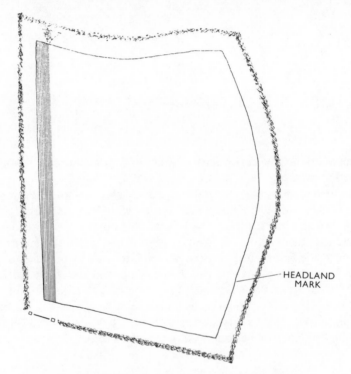

HEADLAND
MARK

Fig. 90. A field marked out for "one-way" ploughing.

Disc Ploughs

The disc plough is a type of plough generally used where conditions are very hard and rough and totally unsuitable for use of the conventional mouldboard plough. It is not a plough often used in countries with temperate and moist climates except perhaps to carry out reclamation work or to plough land in which there is much stone and tree root. In tropical countries where the climate is hot and the land bakes hard, the disc plough is more likely to be in favour.

This plough is unlike the mouldboard plough in that it does not have a share, mouldboard, or coulter to cut and invert a furrow slice, but instead it has a heavy steel concave disc of about 60-70 cm (2-2 ft 6 in.) diameter which carries out a similar but not by any means identical function to the mouldboard and share. As a disc plough is pulled along, the discs rotate, cutting into the soil and the so-called furrow slice is caused to rise up in the concavity of the disc, to be broken up as it does so, and then thrown sideways. Tidy, even, completely inverted furrows are not normally produced, nor are trash and weeds buried as with a conventional plough.

Disc ploughs may be trailed or mounted as are the mouldboard ploughs. A trailed type may have up to six discs and is thus a very heavy implement utilizing its weight to assist penetration when working in hard conditions. A mounted type usually has up to five discs and is thus necessarily lighter to enable it to be carried by a tractor. However, to achieve penetration when working in hard conditions it may be necessary to add weight to the plough. Disc ploughs with more than six discs are usually known as Polydisc ploughs and have discs of about 50 cm (20 in. approx.) diameter. The discs may be plain or cut-away. Cut-away discs, see Fig. 91, more effectively bite and cut into hard soil conditions.

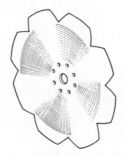

Fig. 91. A cut-away concave disc.

Figure 92 shows an illustration of a typical mounted disc plough. So that the plough can be made to work satisfactorily in various conditions a number of adjustments are usually provided on it. The mounted plough is attached to the tractor on the three-point linkage and adjustment to the top link and/or the right-hand lifting rod serves to level the plough. Depth of work is set by alteration of the height of the depth wheel or where a depth wheel is not fitted, by a setting on the hydraulic depth control. When the depth is set, the beam should be set horizontal by adjusting the top link and the plough level by adjusting the lifting rod.

Width of furrow may in some cases be altered by changing the position of the disc in relation to the beam to which it is attached. This can involve, in effect, slight rotation of the disc so that the angle of the disc to the line of pull is greater or less. Whilst this will alter the furrow width it will also make the plough more difficult to pull and cause the furrow to be thrown farther sideways, if the angle is made greater.

Alteration of the vertical angle of the disc will affect penetration by the disc. The more vertical the disc, the greater will be the penetration.

The rear wheel of a disc plough is spring loaded to maintain a pressure causing it to cut into the soil and it is also set to run at an angle to counteract the considerable side thrust set upon the plough when the furrows are moving over the discs. On some ploughs there may be vertical and lateral adjustment on this wheel to correct any tendency of the plough to swing at the rear.

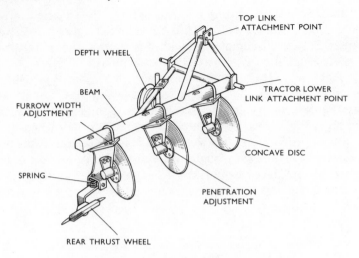

Fig. 92. A three-furrow tractor-mounted disc plough.

Chisel Ploughs

Chisel ploughing is a term that has come into use in recent years to describe a certain method of field cultivation. This operation is not in fact a method of ploughing as is generally understood from the meaning of the word ploughing, but is more of a heavy cultivating operation. When we think of ploughing we usually think of the use of an implement to cut and upturn a furrow to expose the underside of the furrow to the atmosphere. Chisel ploughing consists of pulling heavy cultivator tines through the soil, normally at a depth greater than that at which conventional ploughing would be done, and bursting up the underlying layers of soil without bringing subsoil to the surface. The tines of the implement are sturdy and strong to withstand the stresses applied when they are working at depth where soil conditions are harder. The implement frame is also strongly constructed usually of box-section steel to withstand the heavy loading applied to it. Generally the frame is of all welded structure and allows for alteration of the tine spacing, the tines being clamped and bolted to the beams.

Chisel ploughing is considered to have a number of advantages which many farmers think justify the use of this type of implement. These advantages include bursting up the subsoil, improving drainage, aerating the soil, breaking up a soil pan and pulling up deep-rooted weeds. It is a useful implement to use in hard conditions and in reclaiming of rough land and orchards, etc.

Figure 93 shows an illustration of this type of implement which is generally constructed for use on the three-point linkage of the tractor, and depth of work is controlled by the hydraulic system of the tractor. Because the implement is heavily constructed and has a heavy job to do it requires considerable power to pull it. Something in the region of 60 to 100 h.p. is usually required, depending on the implement size. The implement may be fitted with only nine tines and have a working width of about 3 metres (9 ft. approx.). The tines are up to 60 cm (2 ft) long. Replaceable steel or cast-iron points may be fitted to the tines.

Working of the implement is fairly straightforward, the main thing to remember being that it must be kept level when in work and adjustments for this level are made on

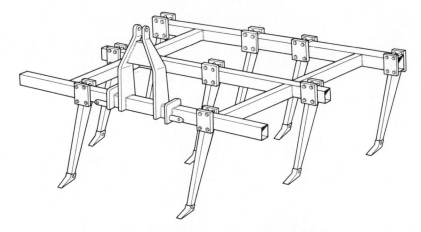

Fig. 93. A heavy duty cultivator or chisel plough.

the tractor top link and/or the right-hand lifting rod of the tractor. If these are correctly set the frame of the implement should be parallel with the back axle of the tractor when looked at from the rear and parallel with the field surface when looked at from the side.

Sub-soilers

The function of the sub-soiler is to penetrate to deeper depths than conventional cultivation machinery and break up the layers of soil which have become compacted due to the movement of heavy machinery or as a result of ploughing. These compacted areas prevent the natural drainage of the soil and also inhibit the passage of air and nutrients through the soil structure.

Sub-soilers consist (Fig. 94) of one or more heavy tines which break through the impervious layer, causing a shatter. Normal depth of work would be approximately 500 mm (20 in.) although this should be related to the depth of the plough pan and at intervals of 1 m. The power required to pull such implements is high and, in order to either reduce the draught or increase their efficiency, variations of the conventional sub-soiler are available.

Vibratory machines improve the amount of shatter by means of a moving share which is driven by the p.t.o.; wings may be attached to the foot of the sub-soiler. These run in lateral fissures and increase the area of soil which is worked considerably, with little increase in power requirement. The operation should be carried out under very dry conditions so that maximum fissuring of the soil is achieved.

Routine Maintenance of the Plough

Routine maintenance of a plough in service will consist mainly of lubrication attention. Generally, each lubrication point will require attention once daily. Some will require oil lubrication and some will require grease. It is often recommended that the disc coulter bearings are greased twice daily.

Replacement of ploughshares will be necessary as work proceeds.

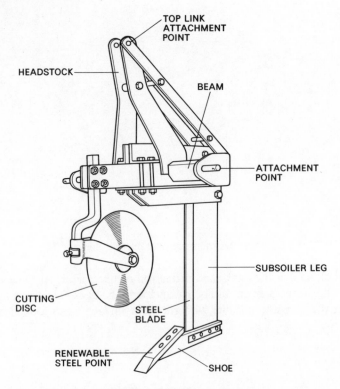

Fig. 94. A typical sub-soiler.

At the end of a day's work the plough mouldboards should be coated with oil to prevent rusting and at the end of the season's work mouldboards and disc coulters should be coated with an anti-rust preparation. This will ensure that the mouldboards keep their "shine" and work properly when ploughing is done again.

As far as disc ploughs are concerned, the same sort of maintenance attention will be required. In some cases it may be found that the disc bearings are prepacked with grease and sealed so that further lubrication is neither necessary nor possible.

CHAPTER 15

Cultivation Machinery

CULTIVATION machines are used to break down the soil before or after a crop is sown, for covering seeds, for consolidating the soil and (with special fittings) for hoeing out weeds. Those capable of deep penetration and having heavy tines are usually classed as cultivators, while more lightly constructed implements for shallow depth and surface working are referred to as harrows. The effects which cultivation machines will have will depend a great deal on the type of soil being worked, and the notes given below on the uses of the machines are a general guide only.

Cultivators

These machines may be attached directly to the tractor three-point linkage or pulled from the tractor drawbar. Working depth may be controlled in three ways.

(a) *Wheel control:* depth-limiting wheels are linked by an adjustable mechanism to the cultivator frame. This method maintains the machine at a very accurately controlled working depth, but little weight is transferred to the rear wheels of the tractor and thus adhesion only slightly assisted. All trailed cultivators use this type of depth control. A fully mounted cultivator is shown in Fig. 95.

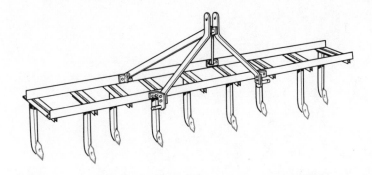

Fig. 95. A mounted cultivator.

(b) *Draught control:* the force set up in the top link of the three-point linkage is fed to the control valve in the hydraulic system; the working principles are exactly the same as those described when dealing with draught control for ploughing depth (see p. 137). Weight is transferred to the rear wheels of the tractor and thus wheel grip is improved, but the working depth tends to vary if the soil texture is not uniform.

(c) *Position control:* in this system the working depth of the implement is limited by a mechanical stop on the tractor or by locking the oil in the hydraulic cylinder. When this method is used, a constant working depth is maintained as long as the field is level.

Effects and Uses of Cultivators

Cultivators break the soil and tend to lift it; unrotted trash is brought to the surface where it can be killed by the sun. They are often used in heavy land to avoid compaction and for preparing land for root crops when a deep tilth is required.

Cultivators can often be adapted for row crop hoeing; special hoe blades are fitted which cut away the weeds and small concave discs can also be attached. The cultivator frame is drilled so that the position of the tines can be altered to suit the row widths of the crop being hoed. It is usual to set the attachments so that the discs run close to the rows in order to clear away the soil from the young plants and to prevent them from being buried by the following L-hoes. The broad A-hoes complete the hoeing in the middle of the rows. When the plants are fairly large, the discs are not usually fitted.

Disc Harrows (Fig. 96)

These comprise a number of sets or gangs of concave discs which can be set at a variable angle to the direction of travel. The greater this angle the more *severe is the treatment of the soil.*

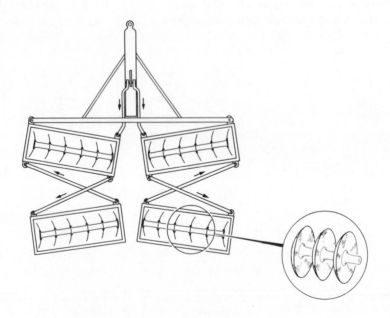

Fig. 96. The disc harrow, showing the principle of alteration of the
angle of the gangs of discs.

Effects and Uses of Disc Harrows

As the discs are drawn along they turn and break down the soil surface; there is a consolidating effect on the subsoil. Discs are very useful in the light land as they cultivate the soil without tending to make it too "fluffy", which a cultivator might do. They do not bring unrotted trash to the surface and are thus used for working recently ploughed grassland or stubble. Disc harrows are sometimes used for stubble cleaning, i.e. the light cultivation of land after a corn crop has been removed; weed seeds are encouraged to germinate and they are killed by the ploughing which is usually done a few weeks later.

Spring-tine Cultivators (Fig. 97 and 98)

The tines are made of spring steel; replaceable steel digging points are fitted. On larger types the tines are mounted on frames similar to those of ordinary cultivators but the smaller models are built into frames similar to those of a harrow. The working depth of the former type is controlled in the same way as a cultivator; but on the lighter types the bars, to which the tines (usually C-shaped) are attached, can be rotated and locked in the desired position. As the bar is turned, the digging points which are fixed to the ends of the tines are forced into or lifted out of the ground.

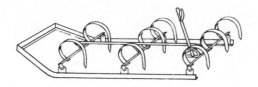

Fig. 97. A spring-tine cultivator.

Effects and Uses of Spring-tine Cultivators

As the machine is drawn through the soil the tines judder, thus helping to break down any clods. Spring-tine cultivators are particularly useful for preparing a seed-bed quickly and also for lifting unrotted trash to the surface, where it can be killed by the sun. They can be used for preparing seed-beds for most arable crops but they are normally limited to a working depth of about 23 cm (9 in. approx.) whereas the rigid tine cultivators can be used for working at greater depths.

Figure 99 shows the action of the common types of cultivator tines.

Seed Harrows (Fig. 100)

The tines are bolted to frames which are usually pulled in sets of three to six by a *harrow pole* assembly. Seed harrows are available which can be attached directly to the three-point linkage of the tractor.

Effects and Uses of Seed Harrows

The tines, which are about 15-23 cm (6-9 in. approx.) long, break down the soil up to their working depth; they have a small consolidating effect on the subsoil. Heavy long-tined harrows are effective in preparing land for drilling, and in some heavy land

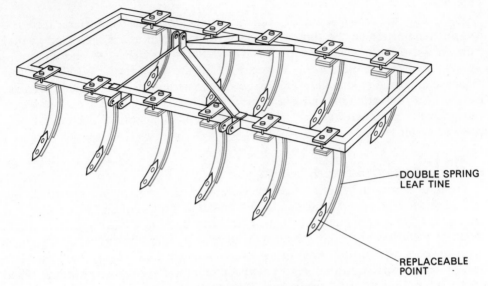

DOUBLE SPRING
LEAF TINE

REPLACEABLE
POINT

Fig. 98. A mounted spring-tine cultivator.

RIGID TINE

FLAT SPRING TINE
VIBRATING ACTION

HEAVY SECTION SPRING
TINE VIBRATING ACTION

Fig. 99. The action of different types of cultivator tines.

areas the tines are curved to provide penetration. Light harrows are generally used to harrow in seed after drilling and also to harrow winter-sown cereal crops in the spring to suppress annual weeds and break the winter soil cap.

Chain Harrows (Fig. 101)

The chain webbing may be plain but is often fitted with renewable double-ended tines; by reversing the harrow either the short or the long tines may be used.

Effects and Uses of Chain Harrows

This type of harrow is widely used for harrowing grassland either to pull out any dead grass and aerate the sward or to spread dung pats. However, some people think that this practice helps to spread parasites. The dead grass collects in the tines and it may be necessary to stop repeatedly to clean the harrow. If the front of the harrow is raised slightly by shortening the tow chain, there is a tendency for the trash to be

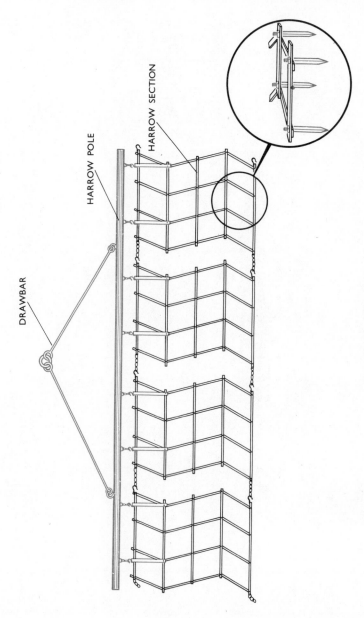

HARROW SECTION

HARROW POLE

DRAWBAR

Fig. 100. Seed harrows.

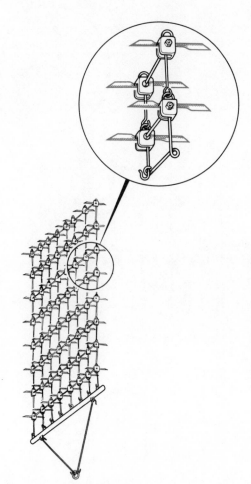

Fig. 101. Chain harrows.

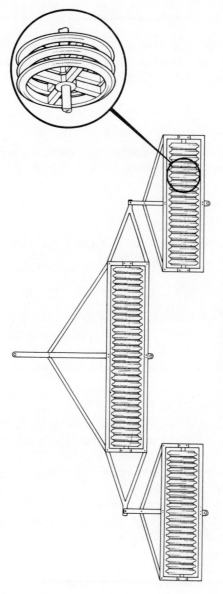

Fig. 102. Cambridge rollers.

formed into rolls and to clear the chain webbing automatically. Chain harrows can also be used for harrowing in seed (providing the field is clear of trash); for harrowing winter-sown corn and for levelling potato baulks after planting.

Cambridge or Ridged Rollers (Fig. 102)

The rollers are built up of narrow cast-iron sections and, due to the thin ridges on the rims of the rollers, this type is very effective for consolidating and breaking down soil.

Uses of Cambridge Rollers

This type of roller is used for clod crushing and consolidation prior to drilling, also for rolling winter-sown cereals in the spring to encourage tillering.

Smooth Rollers

The rollers are usually made of cast iron in two or three sections (to avoid "scuffing" when turning). Smooth rollers have a crushing and consolidating effect but they are not as good for this job as Cambridge rollers.

Uses of Flat Rollers

Flat rollers are particularly useful for rolling root land before drilling; this produces a level surface and a fine compact tilth. The compaction helps to draw to the soil surface the moisture which is essential for good seed germination. They are also used in the spring to roll in stones in fields intended for silage and hay.

Rotary Cultivators (Fig. 103)

These machines are directly mounted to the three-point linkage of the tractor. A p.t.o. shaft takes power from the tractor to a bevel gearbox and the output shaft of the latter drives the rotor through sprockets and chains. The rotor revolves in the same direction as the tractor wheels, and the tines are usually fixed to the rotor shaft in the form of a helix to avoid jerky action.

As the machine is drawn along, the tines break up the soil; the texture of the resulting tilth is controlled by the relationship of the forward to rotor speed. The former is varied by the tractor gearbox, the latter is affected by the selection of gears in the bevel box assembly and/or sprockets in the chain drive assembly. Various types of tine are available for use in different conditions. A safety clutch is usually provided in the p.t.o. drive or the chain drive assembly to prevent damage to the rotor.

Uses of Rotary Cultivators

1. Getting a quick tilth: the effects of ploughing and cultivation can be realized at one pass.
2. Mulching-in surface trash which might clog a plough, e.g. "working-in" straw or bracken eradication.
3. The breaking up of large clods.

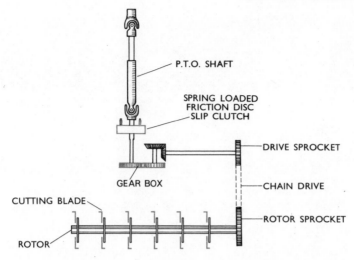

P.T.O. SHAFT

SPRING LOADED
FRICTION DISC
SLIP CLUTCH

DRIVE SPROCKET

GEAR BOX

CHAIN DRIVE

CUTTING BLADE

ROTOR SPROCKET

ROTOR

Fig. 103. Layout of the drive on a tractor rotary cultivator.

Rotary cultivators which work on a vertical axis are also extremely useful in the preparation of a seed-bed although they are less suited to primary cultivations, Fig. 104. Mounted on the tractor hydraulics, the power is transmitted by the p.t.o. to a slip clutch which protects the machine from overloading. From here the drive enters a gearbox which provides a selection of rotor speeds, and on to a series of gears which cause the tines to rotate. Each adjoining set of gears is travelling in the opposite direction of rotation. Less air is incorporated into the tilth by this system of cultivation and consequently the seed-bed is firmer. A crumbler roller may also be fitted which will break surface clods, improve consolidation and level out the seed-bed.

The Power Harrow (Fig. 105)

This implement is widely used for seed-bed preparation. It is fitted with straight vertical tines up to 23 cm (9 in.) in length and these are bolted to straight horizontal tine bars. An implement of this type may have from two to four tine bars fitted, the four-bar implement being used for work on heavy land and the two-bar implement for use on light to medium land.

In work, the tine bars, which receive the drive from the p.t.o. of the tractor, reciprocate sideways at right angles to the line of travel of the tractor, and thus produce the necessary tilth by stirring through and striking clods and earth. Usually on the four-bar implements the extent of the side movement of the tine bars increases from front to rear, for example, the front tine bar may only move about 10 cm (4 in.) sideways, whilst the second bar moves about 20 cm (8 in.), the third 30 cm (12 in.) and the fourth 40 cm (16 in.). This allows for a progressive breaking-down action on the soil. Also, and again to allow for a progressive breaking down of the soil, it is advisable to operate the implement so that it is slightly tilted to the rear. This can be done by lengthening the top link attaching the implement to the tractor and then for a given depth setting, the front tines will always work shallower than the rear. This setting is more important on a four-bar implement.

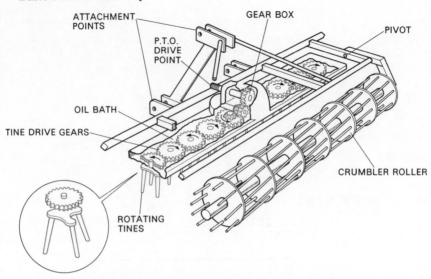

Fig. 104. Working components of a mounted rotary cultivator.

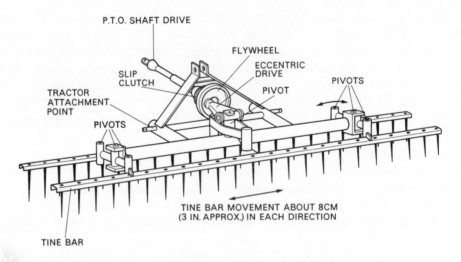

Fig. 105. A power harrow.

Effects and Uses of the Power Harrow

This implement is used mainly for the preparation of a seed-bed, whether in spring or autumn, and in some soils and conditions it can produce in one operation the same tilth as would be produced by three operations with other conventional cultivators and harrows. The tines do not bury soil but stir and strike it down to the depth at which the tines are working, therefore for spring cultivations the winter-weathered soil is soon made into a suitable tilth.

For any given soil conditions the type of tilth produced will depend on the forward speed of the tractor, the reciprocating tine bar movement being constant. Therefore, decreasing the forward speed will produce a finer tilth and increasing the forward speed a coarser tilth. Also, of course by increasing or decreasing the p.t.o. speed, thus varying the speed of reciprocation of the tine bars, will vary the condition of the tilth produced.

Being a power-driven implement it is necessary to protect it against damage should the tines strike obstructions, therefore a slip clutch is fitted in the drive line to the tine bars. The clutch must be correctly adjusted. Also a gearbox and greasing points on the implement require periodic attention.

Dutch Harrows

These simple machines are very popular in arable areas and in use they achieve much the same as the previous implement. They are primarily for use on light soils and the operation may require two or more passes but the completed work can provide an ideal surface on which to drill small seeds. Mounted on the three-point linkage, a sturdy frame supports several rows of steel tines each held individually in position by a lock bolt, which also provides an adjustment for the working depth of the tine (Fig. 106). A sturdy wooden beam mounted at an angle across the front of the implement provides a levelling effect and again a crumbler roller on the rear improves the finished tilth.

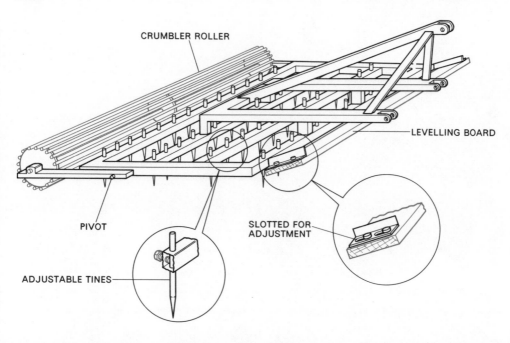

CRUMBLER ROLLER

LEVELLING BOARD

PIVOT

SLOTTED FOR ADJUSTMENT

ADJUSTABLE TINES

Fig. 106. Dutch type levelling harrow.

Routine Maintenance of Cultivating Machinery

The type of maintenance done to cultivating machinery depends on the type of implement. Cultivators are not normally complicated implements and maintenance will consist mainly of replacing or reversing the points. The points must not be worn down to the stage where the ends of the tines also start wearing otherwise new points may not fit properly.

If a cultivator is fitted with wheels, the wheel bearings will require daily lubrication.

Rollers and disc harrows are fitted with greaser points that require daily attention when in use. A tractor rotavator has a number of grease points requiring daily attention and generally also gear housings in which the gears run in oil. These should be kept topped up with oil of the correct grade and it is usually recommended that the oil is drained off and replaced with new oil, annually.

CHAPTER 16

Corn Drills

IN A typical corn drill the seed-box is carried on a frame supported by wheels from which the drive to the seed-metering mechanism is taken. The drive may take the form of gear and/or chain and sprockets but the design always allows the overall ratio to be altered so that seeding rates can be varied. When the seed has been passed through the metering mechanism it falls down a number of plastic, rubber or metal tubes into the coulter assembly. The purpose of the latter is to cut a groove in the soil, into which the seed can fall.

Seed-metering Mechanisms

1. *Internal Force Feed (Fig. 107a)*

A number of metering units, about twelve to fifteen is usual, are attached to the underside of the seed-box; they are driven by a shaft passing through the feed discs.

The feed disc has teeth on the inside of the rim; the pitch of these teeth is different on each side of the disc and a flap or *blanking plate* allows the seed to be diverted to either side of the feed disc.

As each disc is turned, seed is carried between the teeth from the seed-box to the feed outlet, where it falls down the feed tube to the coulter.

FEED ADJUSTMENTS

1. The position of the flap or blanking plate which diverts the seed to a particular side of the feed disc.
2. The speed of rotation of the feed disc; this is controlled by a gearbox assembly in the drive or by changing gears or sprockets in the line of drive.
3. Special restrictors are often available; when fitted in the unit these attachments restrict the feed outlet and thus cut down the seeding rate.

2. *External Force-feed Type A (Fig. 107c)*

The units which are made up of a fluted roller running in a hollow casting are driven by a shaft running the length of the drill. As the roller turns seed is carried in the flutes or grooves from the seed-box to the feed outlet. The roller can be slid sideways out of the unit when a stationary blanking plate is substituted; thus a lesser amount of feed roller is exposed to the seed. On some machines the bottoms of the units can be pivoted to allow a variation in feed outlet size, this also allows adjustment for different sizes of seeds.

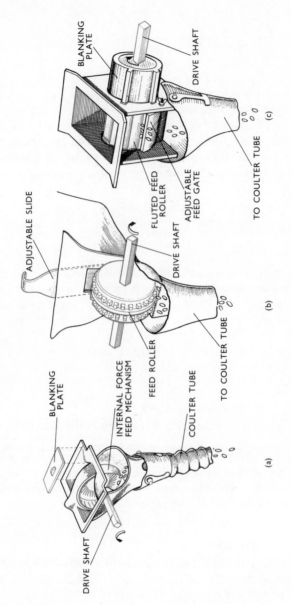

Fig. 107a. Internal force feed.

Fig. 107b. External force feed (toothed roller).

Fig. 107c. External force feed (fluted roller).

FEED ADJUSTMENTS

1. The speed of rotation of the fluted roller by gearbox adjustment or the substitution of gears or sprockets in the line of drive.
2. The position of the fluted roller.
3. The position of the bottom flap which controls the feed outlet size.

3. *External Force-feed Type B (Fig. 107b)*

This is a development of the type of external force feed described above. The seed flows past an adjustable slide to the feed roller, which is placed at the rear of the seed-box. The bottom plate under the feed roller is adjustable.

FEED ADJUSTMENTS

1. Speed of rotation of feed roller by gearbox adjustments or the substitution of gears or sprockets.
2. The position of the adjustable slide.
3. The position of the bottom plate.
4. The use of feed rollers with one or two rows of teeth.

Depth of Sowing

The depth of sowing depends on the position of the coulters. These are carried on arms attached to a bar at the front of the drill frame which are spring-loaded downwards: the pressure on the springs is applied by a bar connected to the depth-control handle. The spring pressure can also be altered on each coulter independently. On some modern drills the position of the carrying wheels and thus the depth of sowing can be altered hydraulically.

Types of Coulters

Disc

Disc coulters are very suitable for most conditions in Great Britain. They do some cultivation during the furrow-opening process and so the land does not need to be worked quite as much as with Suffolk coulters. The discs tend to cut through any surface trash but in stony ground they may become jammed. Disc coulters do not maintain as straight a path as the Suffolk type and they are thus not so good for drilling root crops which must be hoed at a later date. They are more expensive initially than Suffolk coulters but do not wear out as fast. The bearings must be greased regularly or they may seize up and "flats" may be worn on the discs.

Suffolk

Suffolk coulters are particularly useful for drilling root crops; due to their "fin-type" shape they keep very straight so that hoeing of the crop at a later date is made as easy as possible. They wear out faster than disc coulters because of the soil abrasion. Quite often the coulters which run behind the tractor wheels will, due to soil compaction, wear out sooner than the rest of the set. Suffolk coulters are effective in stony ground as there are no moving parts which can become jammed.

Combine Drills

Many modern drills are of the *combine* type: a separate box for fertilizer is fitted to the drill frame. Metering units, fitted to the bottom of this box, feed the fertilizer either into a common outlet with the seed or down separate tubes to the coulter assembly. The fertilizer metering units are driven through gears and/or chains and sprockets from the drill wheels.

Fertilizer Metering Unit (Fig. 108)

This is often called the *star wheel feed*. Fertilizer is carried by the star wheels under an adjustable slide into a compartment which lies behind a partition. There is a hole in the bottom of this compartment through which the fertilizer is swept, the spaces between the fingers being kept clear mechanically. The shape of the slide housing pushes fertilizer, on the solid portion of the star, over to the finger spaces.

FEED ADJUSTMENTS

1. Speed of rotation of the star wheels: this is usually varied by changing gears or sprockets in the line of drive.
2. Position of the adjustable slide: these slides are controlled by a shaft passing through the fertilizer box. A handle attached to one end of this shaft is retained in any particular position by a quadrant on the end of the box.

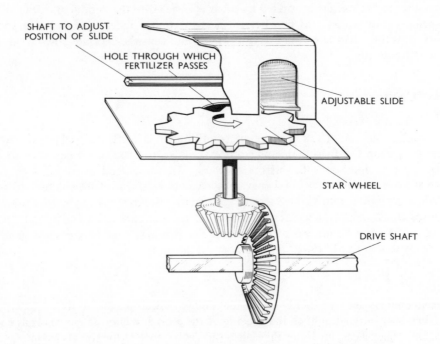

Fig. 108. A fertilizer metering unit on a combine drill.

Markers (Fig. 109)

Markers are often fitted to drills to help the operator to drive so that the drill runs are properly spaced.

Markers can be set thus:

1. Measure the front wheels of the tractor. Let this distance be *A*.
2. Measure the distance between the outside coulters on the drill. Let this be *B*.
3. Set each marker beyond each outside coulter by the following distance:

$$\frac{B - A}{2} \text{ plus 1 row width.}$$

The tractor is steered so that one front wheel follows the marker furrow made on the previous run.

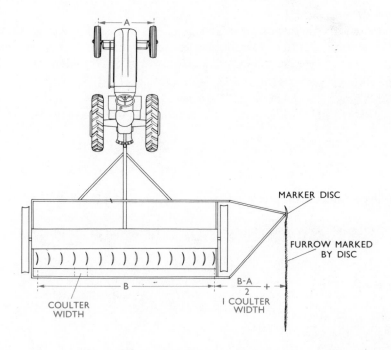

Fig. 109. Marker setting on a grain drill.

Checking the Sowing Rate (Calibration)

It is possible to check whether or not a grain drill is sowing the amount of seed, or fertilizer also in the case of a combine drill, that it should sow at a given setting.
Possible reasons for doing this check are:

(a) The drill setting instructions may have been mislaid.
(b) Wear on the sowing mechanisms is affecting sowing rate: this may be particularly so on the fertilizer side of a combine drill.
(c) A setting may not be known for a particular type of seed being sown, i.e. mixed corn.

In addition to the above if one considers that a grain drill, due to some reason or other, may be sowing a heavier seeding rate than it is set to sow, say 10 kg per hectare (9 lb/acre approx.) more, whilst this may appear to be a negligible amount on 1 hectare or acre, it can amount to a considerable quantity over-sown on a large area. On 400 hectares (1000 acres) the amount of seed over-sown would be 4000 kg (9000 lb). It should be noted that different samples of the same variety of a given seed can sow at a different rate for the same setting on a drill. This is often due to the seed being of a different size. Moisture will also affect seed rate along with such factors as whether or not the seed is dressed. Checking the sowing rate of a grain drill is referred to as calibration and there are two methods of doing this. One is known as the field method, and the other is a static calibration. Both of them can be applied to both sides of a combine drill.

The Field Method

To check the sowing rate of the grain side of the drill, set the sowing mechanism to the rate to be checked, if the setting instructions are available. If they are not available a setting will have to be guessed; in such a case it will be better to do a static calibration.

Fill the grain hopper to the top and level.

Sow an area equal to ·04 hectare (1/10 acre).

The distance to travel with the drill to cover this area will be as follows:

$$\frac{404}{\text{Sowing width of drill in metres}} \qquad \text{or} \qquad \frac{484 \times 9}{\text{Sowing width of drill in feet}}$$

EXAMPLE (a) Sowing width of drill

$$= 3 \text{ metres}$$

$$\therefore \quad \text{Distance to travel}$$

$$= \frac{404}{3} = \frac{135 \text{ metres}}{} \text{ approx.}$$

EXAMPLE (b) Sowing width of drill

$$= 12 \text{ ft,}$$

$$\therefore \quad \text{Distance to travel}$$

$$= 484 \times 3 = 121 \times 3$$

$$= 363 \text{ ft} = 121 \text{ yd.}$$

After sowing this area, measure the quantity of grain required to fill the grain hopper to the top and level, as it was at the commencement of drilling. This quantity, multiplied by 25 will give the amount being sown per hectare or multiplied by 10 the amount being sown per acre.

If it is not desired to actually sow the seed during this field calibration, the coulter tubes can be removed from the coulters to hang freely but with small polythene bags tied to their ends to collect the seed. After travelling the required distance, the total amount of seed delivered is collected, weighed and multiplied by 25 or 10 as described above to give the result on a hectare or acreage basis.

The Static Method

The advantage of doing a field calibration is that the check is being carried out under the conditions in which the drill will work. This is not so in a static calibration, however, allowances can be made for differences between the two methods, the main one of which is the occurrence of wheel-slip which takes place in actual field drilling. This can amount to 10% or more according to conditions. When wheelslip does occur it means that the drive wheel turns more times than it should over a given distance. Therefore, the sowing mechanism will also turn more times over the same distance and more seed will be sown than ought to be.

If wheel skidding takes place, it will have the reverse effect on the sowing.

To carry out a stationary calibration, proceed as follows:

(a) Jack up the driving wheel of the drill.

(b) Remove the coulter tubes from the coulters and put the ends into containers, e.g. buckets, etc., three or four may feed into one container.

(c) Put a small quantity of seed in the hopper, say 10 or 12 kg (22 to 25 lb).

(d) Set the drill at the setting to be checked.

(e) Put the drill into gear and turn the driving wheel a number of times equivalent to travelling over ·04 of a hectare (one-tenth of an acre).

EXAMPLE (a)   If the drill is 3 metres wide it will cover the area of 1 hectare

when it travels $\dfrac{10,000}{3}$ metres = 3333 metres.

Note: 1 hectare = 10,000 square metres approximately;

∴ if ·04 (1/25) of a hectare has to be covered, the distance travelled will be

$\dfrac{3333}{25}$ = 133 metres approximately;

∴ if the driving wheel of the drill is, say, 1·25 metres diameter, the number of turns required to travel the 133 metres will be

$$\frac{133}{1\cdot25 \times \pi} = \frac{133}{1\cdot25 \times 3\cdot14} = \frac{133}{3\cdot9} = \underline{34} \text{ approximately.}$$

Note: Take π as 3·14.

EXAMPLE (b)   If the drill is 9 ft wide it will cover the area of 1 acre when it travels

$\dfrac{4840}{3}$ yards = 1613 yards approximately.

Note: 1 acre = 4840 square yards;

∴ if one-tenth of an acre has to be covered, the distance travelled will be

$\dfrac{1613}{10}$ = 161 yards approximately;

∴ if the driving wheel of the drill is, say, 4 ft diameter, the number of turns required to travel the 161 yd will be

$$\frac{161}{1\cdot3 \text{ yd} \times 3\cdot14} = \frac{161}{4} = \underline{40} \text{ approx.}$$

EXAMPLE (c)   In this example the formula $\dfrac{1400}{D \times S}$ may be used to achieve the same result. The letters D and S refer to diameter of the wheel in FEET and the sowing width of the drill in FEET.

To apply this to the figures in Example (b)

$$\frac{1400}{D \times S} = \frac{1400}{4 \times 9} = \frac{1400}{36} = \underline{39 \text{ approx}}.$$

There is a slight difference in the answer but throughout the example approximations have been used and for practical purposes of calibration would produce a reasonable result.

After turning the driving wheel the required number of turns, the seed which is sown into the containers is collected, weighed and multiplied by 10 or 25 according to whether Metric or English figures and calculations have been used. The resultant answer will be the amount of seed that the drill should sow at that particular setting.

This calibration result has been unaffected by field conditions. Therefore, it may be sensible to allow for a percentage of wheelslip or skidding.

When carrying out this static calibration it is possible to do it by collecting the seed from only one coulter tube or in fact any number less than the total number. If the drill being calibrated was, say, a fifteen-coulter drill, and the seed was collected from only one coulter, it would be necessary to multiply the weight of seed taken from this coulter by 15 then by 10 to give the acreage sowing. In effect, the weight is multiplied by 150. The use of this method does not alter the initial calculation to find out how many times to rotate the driving wheel, nor does it mean that the formula cannot be used.

The main disadvantage of using a single seeding unit is that should this particular unit be sowing either lighter or heavier than any of the others, the discrepancy is multiplied by 150 when the calibration is completed.

It is probably better to use all the seeding units if possible when calibrating in this way to get a more accurate result of the overall sowing rate of the drill.

When it has been determined by the calibration how much seed, or fertilizer in the case of the combine drill, is being sown to the acre for a particular setting, a note should be made of the setting for future reference. In fact the best way of ensuring that the setting details do not get lost is to paint the details on the inside of the grain hopper lid.

On some modern drills calibration is assisted by having small trays provided which can be attached in place of a number of coulter tubes to catch the metered seed; and where, as with some drills, it may be necessary to jack up and turn the land wheel to calibrate, most manufacturers now provide a crank handle which can be inserted into the drive mechanism and turning it a specified number of turns will simulate the drilling of one-tenth hectare or acre as the case may be.

To carry out a static calibration in this case proceed as follows:

(a) Remove the coulter tubes from the coulters and attach the calibration tray to the metering units.
(b) Put a small quantity of seed in the hopper, say 10-12 kg (22-24 lb).
(c) Set the drill at the setting to be checked.

(d) Insert crank handle and turn the drive mechanism the number of times equivalent to travelling of ·1 hectare.

In the case of drills not fitted with this facility to turn the drive mechanism it will be necessary to turn the land wheel. The number of turns can be determined as follows.

Example

If the drill is 4 m wide it will cover the area of 1 ha when it travels 2500 m.

Note: 1 hectare = 10,000 m².

∴ If 0·1 (1/10) of a hectare has to be covered the distance travelled will be

$$\frac{2500}{10} = 250 \text{ m.}$$

∴ If the driving wheel of the drill is, say, 1·25 m diameter the number of turns required to travel 250 m will be:

$$\frac{250}{1·25 \times \pi} = \frac{250}{1·25 \times 3·14} = \frac{250}{3·93} = 63·3 \text{ turns.}$$

Note: Take π as 3·14.

(e) The seed collected in the calibration tray can now be weighed.

(f) Divide the weight of seed by the number of metering units which fed the collection tray and then multiply by the total number of coulters. This figure then represents the application rate for 1/10 ha, ∴ × 10 for rate per hectare.

When it has been determined how much seed is being sown to the hectare for a particular setting a note should be made for future reference.

Direct Drills

Direct drills have been introduced to speed up the drilling operation by reducing, or in some cases eliminating, the need for previous cultivations. The drill is very similar to the conventional grain drill except that it needs to be of heavier construction to ensure that the coulters will penetrate the ground. The furrow into which the seed are dropped is formed by a "triple-disc" system, Fig. 110. A single carbon steel disc cuts a slit into the uncultivated soil. Immediately behind the single disc run two discs which are angled so that a "V" slit is opened in the soil into which seed is delivered. Even depth of drilling is maintained by the weight of the drill pressing the discs into the ground through a rubber spring which regulates the load on each disc assembly.

Routine Maintenance to Grain Drills

There may be as many as forty or more lubrication points on a combine grain and fertilizer drill. Almost all of these will be greasing points and will require daily attention. Disc coulters in particular should receive regular attention. Chains and sprockets may require daily oiling but they should also be kept as clean as possible and free from dust and grit otherwise they may wear rapidly. Clutches for putting the various mechanisms in and out of gear should also be kept well oiled and free from dirt.

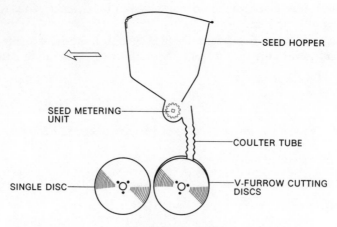

Fig. 110. Direct drill.

The exact detail of lubrication requirements and the location of the lubrication points can only be got from the instruction book applicable to the drill.

At the end of a day's work when drilling grain, it is not entirely necessary to empty the grain hopper if drilling is to be continued on the next day or within a few days. If, however, it is a combine drill that is being used and fertilizer is also being sown, it is most important that the fertilizer hopper is emptied at the end of each day's work. If this is not done, the fertilizer left in the hopper will absorb moisture from the damp night air and the result could be a nasty sticky mess in the hopper that has to be cleaned out entirely, before work can proceed the next day. At the end of each day's work cover the drill completely with a tarpaulin.

At the end of a season's work, thorough cleaning of the whole drill is required. Grain should not be left in the hopper, otherwise it will germinate and grow. If fertilizer is left in, it will corrode the metal components. The following end of season attention should be given:

(a) Remove all grain and brush out the hopper.
(b) Remove all fertilizer and if possible sowing mechanisms within the fertilizer hopper. Other detachable metal parts should also be removed.
(c) Thoroughly scrub out the hopper to remove all traces of fertilizer. Use plenty of water if necessary.
(d) Thoroughly scrub all the fertilizer sowing mechanisms and other metal parts in contact with fertilizer.
(e) Remove any soil, grit or dirt, etc., from external parts of the drill.
(f) Lubricate all grease and oil points.
(g) Brush oil or anti-rust preparation over all metal parts of the drill.
(h) Store the drill under cover.

CHAPTER 17

Fertilizer Distributors

Fertilizer distributors may be divided broadly into three types:

1. Full-width machines which are designed to apply material over a bout width which is as wide as the machine itself; these machines may be trailed or fully mounted.

2. Broadcast machines which apply material over considerably greater distances than their own width. Usually this is 6-7 m but may in some instances be as great as 12·5 m to accommodate compatibility with "tramline" systems. Included in this category are the pneumatic type machines for, although they are not of the broadcast type, they do apply fertilizer over bout widths suitable for agricultural applications on a field scale.

3. Liquid fertilizer applicators which are designed for the placement of fertilizer either in a liquid or gaseous state into the soil. With these machines high amounts of fertilizer may be applied in a single application, giving a progressive release of nutrients through the growing season. Their rate of work is considerably slower than conventional applicators but other advantages may outweigh this drawback.

Full-width Machines

These types of machine are primarily used where it is required to apply high quantities of fertilizer uniformly, for instance vegetable-growing and root-crop production. Although fully mounted on to the three-point linkage they are land-wheel driven and consequently the application rate is not governed by forward speed, as an increase in tractor speed also increases the land wheel drive so delivery rate/hectare remains the same.

Drive from the land wheels passes through either a chain and sprocket system or a gearbox which provide the necessary variations in application rates. A number of different methods of metering may be employed; two, which are the star wheel type and the external force feed type, have already been described in Chapter 16. In all cases the fertilizer is stored in a hopper of approximately 3 m in width with a holding capacity of approximately 500 kg. An extremely simple system is to employ a cushioned pair of rubber rollers (Fig. 111) which are driven in opposite directions to each other so that fertilizer is drawn between the two rollers from where it falls to the ground. Cleaning and maintenance of these machines is made very easy by the absence of numerous moving parts. With full-width machines bout matching is also made easy as it is only necessary to follow the "wheelings" of the previous bout.

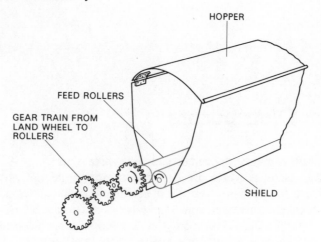

Fig. 111. The gear drive and mechanism of a full width roller
feed fertilizer distributor.

Broadcast Machines

These machines are favoured these days due to the improvement of both machines
and fertilizers. They are inexpensive, easy to clean and permit very high work rates,
particularly with light dressings.

(a) *Spinning Disc Machines (Fig. 112)*

The fertilizer is contained in a hopper which on the smaller mounted machine will
hold 400-500 kg; on larger trailer machines this may increase to 2 tonnes.

With mounted machines the fertilizer is allowed to fall from the hopper on to a
spinning horizontal disc which spreads it over a width of about 6-9 m (20-30 ft approx.).
The disc is driven by a land wheel, from the tractor p.t.o. or by a hydraulic motor
powered by the tractor hydraulic system. The fertilizer is not spread evenly over the
whole sowing width and to achieve an even distribution it is usual to overlap the sowings.
A common practice is to drive so that the fertilizer sown just reaches the wheel marks
made on the previous run.

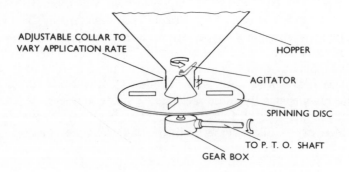

Fig. 112. Spinning-disc fertilizer distributor.

ADJUSTMENT OF APPLICATION RATE

The following factors affect the application rate:

1. *The feed rate of fertilizer from the hopper on to the spinning disc.* This is controlled by an adjustable slide in the bottom of the hopper, an adjustable collar over the hopper outlet, or by moving the disc vertically to allow a greater or lesser flow of fertilizer from the hopper.
2. *The speed of rotation of the spinning disc.* The greater this speed, the greater is the sowing width and thus the smaller the application rate.
3. *The forward speed of the tractor.* The greater the forward speed the smaller is the application rate.

Charts are supplied for each make of machine which indicate the application rates at various p.t.o. and forward speeds and feed slide openings. These provide a good guide to the application rate; but since particular samples of fertilizers and seed vary in texture, the operator may need to make minor modifications to the recommended settings.

Both the plate and flicker (Fig. 115) and spinning-disc type (Fig. 112) of fertilizer distributor are suitable for sowing grass and other small seeds. When grass seed mixtures are sown the disc speed and thus the spread should be reduced: this cuts down separation of different types of seeds due to their different shapes and densities.

CHECKING THE APPLICATION RATE (CALIBRATION)

It has already been mentioned that three main factors affect the application rate of a spinning-disc type of fertilizer distributor, namely, setting of the feed gate on the hopper, speed of rotation of the disc and forward speed. Generally a manufacturer of machines of this type will recommend that a fixed p.t.o. speed, say 540 r.p.m., is used for all work and if different application rates are required this is achieved by altering either the feed-gate setting or the forward speed of the tractor.

With a fixed p.t.o. speed, and applying a given type of fertilizer, the spreading width of the implement is likely to be constant although it will be necessary to alter the machine to suit different types of fertilizer (i.e. prilled, granular, powder) as the flow characteristics of the materials varies and unless provision is made for this the spread pattern will not remain symmetrical.

THE FIELD METHOD

This is probably the best and most practical way of checking the application rate of this type of distributor. The object of doing the check is the same as for the wheel-driven distributor or a grain drill.

Proceed as follows:

(a) Ensure that the distributor is correctly attached to the tractor and with p.t.o. shaft correctly coupled. The distributor spinning disc should be level.
(b) Put a quantity of fertilizer, say 50 kg (1 cwt approx.), into the hopper, spread it level across the hopper and mark where it comes to.
(c) Set the feed gate.
(d) From the instruction book, if available, establish the width of spread likely to be produced for the type of fertilizer being used at the recommended p.t.o. speed.

If the book is not available this width will have to be established by doing a short run, say 15 m (16.5 yd approx.), with the spreader working.

(e) Select the suitable forward gear which, combined with a suitable throttle setting, will give a forward speed of 4 to 8 km/hr (4-5 m.p.h. approx.).

(f) At this speed and setting, travel a distance equal to ·04 (1/25) hectare (1/10 acre). This distance will be equal to:

$$\frac{484}{\text{Width of spread in yards}} \text{ yards } \quad \text{or} \quad \frac{484 \times 9}{\text{Width of spread in feet}} \text{ft.}$$

EXAMPLE (a)   Width of spread: 8 m.

$$\text{Distance to travel} = \frac{404 \cdot 64}{8} = 50 \cdot 58 \text{ m.}$$

EXAMPLE (b)   Width of spread: 8 yd.

$$\text{Distance to travel} = \frac{484}{8} = 60 \cdot 5 \text{ yd.}$$

EXAMPLE (c)   Width of spread: 24 ft.

$$\text{Distance to travel} = \frac{484 \times 9}{24} = \frac{121 \times 3}{2}$$

$$= 181 \frac{1}{2} \text{ ft} = 60 \cdot 5 \text{ yd.}$$

(g) Refill the hopper to the level mark but weighing the quantity of fertilizer required to do this. The quantity required multiplied by 25 or 10 depending on whether Metric or English measurements are used will give the application rate per hectare or acre.

If the implement hopper is graduated on the inside to show quantities of fertilizer in the hopper at different levels it can be helpful when carrying out such a field check.

The following chart is a guide to the quantities of fertilizer that are sown at various application rates when spread on ·04 (1/25) ha (1/10 acre).

125·5 kg per hectare = 5·2 kg (1 cwt per acre = 11¼ lb).
251 kg per hectare = 10·4 kg (2 cwt per acre = 22½ lb).
376·5 kg per hectare = 15·6 kg (3 cwt per acre = 33¾ lb).
502·   kg per hectare = 20·4 kg (4 cwt per acre = 45 lb).
627·5 kg per hectare = 26·0 kg (5 cwt per acre = 56¼ lb).

(b) *Oscillating Spout Machines (Fig. 113)*

Here the fertilizer is allowed to fall through an adjustable aperture, which varies application rate, into a conical-shaped spout which is oscillating at 540 cycles/minute driven off the tractor p.t.o. Centrifugal force spreads the fertilizer in a similar manner to that of the spinning disc. However, no adjustment is necessary to compensate for different types of material.

(c) *Trailed Machines (Fig. 114)*

The spinning disc is driven at a constant speed by the tractors hydraulic system consequently because the flow capacities of different tractor manufacturers vary it is necessary to provide a flow restrictor so that disc speed may be adjusted. Too high a

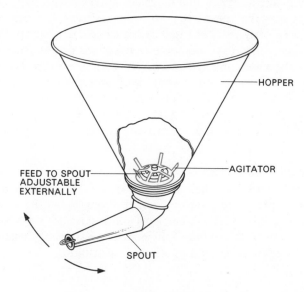

Fig. 113. An oscillating spout distributor.

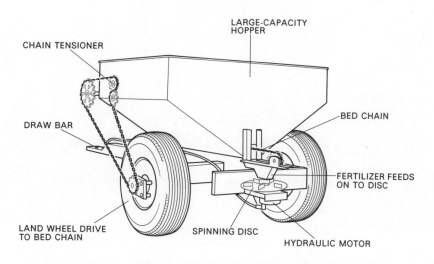

Fig. 114. A trailed spinning disc type fertilizer distributor.

disc speed will result in the fertilizer being thrown too far, giving striping at the overlap point, whereas too low a disc speed will limit the bout width and again cause striping.

Fertilizer is supplied to the disc by a moving-bed chain in the base of the hopper which is driven by a system of chain and/or belt drives from the land wheel. Application rate is varied by altering the speed of the bed chain by using interchangeable sprockets and pulleys and here again forward speed has no affect on the rate applied.

(d) *Pneumatic Machines*

In order to provide the farmer with a machine which will spread as evenly as a full-width machine but with the work-rate of a broadcast machine, pneumatic fertilizer distributors have been designed. The material is stored in a hopper from where it is metered by a pegged roller external force-feed system driven from a land wheel. Fertilizer falls from the metering system into an airstream which is created by a fan operated by the tractor p.t.o., and carried along pipes to nozzle outlets positioned along the lengths of the booms. These machines are only suitable for the application of fertilizer as seeds could receive mechanical damage. However, their design ensures even application on sloping ground or when used in windy conditions.

(e) *Plate and Flicker Machines (Fig. 115)*

Dished plates are fitted horizontally at the bottom of the fertilizer box from which they protrude. These plates are driven by gearing from one of the land wheels of the distributor. As they turn, fertilizer is carried from inside the box past an adjustable slide. The fertilizer is now flicked out of the plates by a series of flickers on a rotating shaft which is driven by the other land wheel of the distributor.

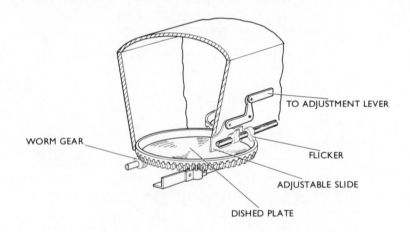

Fig. 115. Rotating plate and flicker fertilizer unit.

ADJUSTMENT OF APPLICATION RATE

1. Speed of rotation of the plates. This is controlled by a gearbox or by changing gears in the line of drive.
2. Position of the adjustable slide. This is controlled by a rod, fixed to the fertilizer box, to which all the slides are attached. Movement of this rod moves all the slides equally.

   Equal amounts of fertilizer can be spread using various combinations of plate speed and gate opening. A low plate speed and a large slide opening is advisable when using damp and lumpy fertilizer, a high plate speed and a small slide opening should be used if the fertilizer is dry and free flowing.

CHECKING THE APPLICATION RATE (CALIBRATION)

Calibration of a grain drill has already been dealt with. The same method can be applied to a wheel-driven fertilizer distributor; the calculations to find the number of wheel rotations is done in the same way and the formula

$$\frac{1400}{D \times S}$$

can also be used. In fact this method of calibration can be applied to various types of wheel-driven seeders or applicators.

As far as the fertilizer distributor is concerned it should be noted, as indeed it should be with a seed drill, that manufacturer's sowing rates at particular settings can only be approximate because of the variations in seed size and fertilizer granule size. Fertilizers in particular can vary considerably in the way in which they will flow through a machine.

There is probably more need to check the application rate of a fertilizer distributor than a grain drill, not only because of the variations in fertilizer but because of the corrosive effects which the fertilizers have on the implement parts. Wear and corrosion to such parts as feed gates, slides, lever linkages, etc., result in inaccurate settings which can do no other than affect the application rate.

Calibration of a wheel-driven fertilizer distributor is carried out as follows:

(a) Jack up the driving wheel of the distributor. In the case of a plate and flicker distributor this may mean both wheels if one is driving the plates and the other driving the flickers.

(b) Set the feed mechanism to the setting to be checked. As with the grain drill, if the setting chart is not available a setting will have to be applied and if necessary altered for a further calibration.

(c) Put a quantity of fertilizer in the hopper. This must, of course, be the type of fertilizer which is to be applied after the calibration is completed.

(d) Spread a sheet of some sort beneath the distributor and covering the spreading area of the implement. A clean concrete floor area, if available, will eliminate the use of a sheet.

(e) Put the mechanism into gear and turn the driving wheel the required number of times to cover ·04 (1/25) ha (1/10 acre).

If a plate and flicker distributor with mechanisms driven from both wheels, then both wheels will have to be rotated and at the same speed. See that the fronts of the plates are full of fertilizer before actually doing the calibration.

(f) Sweep up the fertilizer spread, weigh it, multiply by 25 or 10 according to whether Metric or English figures and calculations have been used. The resultant answer will be the amount of fertilizer that the distributor should apply per hectare or acre at that particular setting.

Examples of calculating the number of wheel rotations required are as follows:

EXAMPLE (a)   A wheel-driven fertilizer distributor with a distribution spread of 5 m and having a land wheel diameter of 1 m is to be calculated. How many times is it necessary to rotate the driving wheel to cover an area equal to ·04 (1/25) of a hectare?

If the distribution spread is 5 m wide it will cover an area of 1 ha when it travels

$$\frac{10,000}{5}m = 2000\,m.$$

Note: 1 hectare = 10,000 square metres.

∴   If ·04 (1/25) of a hectare is to be covered, the distance travelled will be

$$\frac{2000}{25} = 80.$$

∴ If the driving wheel of the distributor is 1 m diameter, the number of turns required to travel the 80 m will be

$$\frac{80}{1 \times 3\cdot14} = 25\cdot4,\ \text{say}\ \underline{25}\ \text{approx.}$$

Note: Take π as 3·14

EXAMPLE (b)   A wheel-driven fertilizer distributor with a spread of 12 ft and having a land wheel diameter of 3 ft 6 in. is to be calibrated. How many times is it necessary to rotate the driving wheel to cover an area equal to one-tenth of an acre?

If the distributor spread is 4 yd wide it will cover an area of 1 acre when it travels

$$\frac{4840}{4}\,yd = 1210\,yd.$$

∴ If 1/10 of an acre is to be covered, the distance travelled will be

$$\frac{1210}{10} = 121\,yd.$$

∴ If the driving wheel is 3·5 ft diameter the number of turns required to travel the 121 yd will be

$$\frac{121 \times 3}{3\cdot5 \times 3\cdot14} = \frac{363}{10\cdot99} = \underline{33}\ \text{approximately.}$$

EXAMPLE (c) Applying the formula $\dfrac{1400}{D \times S}$

$$\frac{1400}{3\cdot5 \times 12} = \frac{1400}{42} = \underline{33}\ \text{approximately.}$$

THE FIELD METHOD

This can be done in exactly the same way as with a grain or combine drill, except that the fertilizer must be spread on the field, see under Checking the Sowing Rate of a grain drill.

Routine Maintenance of Fertilizer Distributors

Good maintenance to a fertilizer distributor is most important if it is to operate satisfactorily and last a reasonable number of years. Fertilizers are very corrosive to metals and a distributor that is not properly cared for can have a very short working life indeed.

Routine maintenance is similar to that carried out on the fertilizer side of a combined drill, see p. 169. The following points need emphasizing.

(a) Lubrication points require daily attention whilst the machine is in use.
(b) Never under any circumstances leave fertilizer in the hopper at the end of a day's work. Empty it and cover it with a tarpaulin.
(c) At the end of the season, remove all traces of fertilizer from the machine. A hose pipe and plenty of water will help considerably when doing this job.
(d) Lubricate all grease and oil points.
(e) Give all metal parts a coat of anti-rust preparation.
(f) Store the distributor under cover.

CHAPTER 18

Farmyard Manure Spreaders

A VARIETY of machines are available for both the loading of manure and the spreading of it. Generally, loading is carried out by the use of a fore-end loader which is hydraulically operated, but other methods of loading can be used. For example, engine-driven elevators used in conjunction with a mechanically moved fork can be used, and there is also a rear-mounted hydraulically operated loader.

The fore-loader can be found on most farms where it can be used for many lifting jobs other than loading farmyard manure. The arms of the loader can be fitted with various attachments for use in other types of work; for example, a bucket or scoop can be fitted for moving soil, sand, gravel and the like. Hedge-cutters can also be attached to these arms and the height at which they cut is set by the hydraulic controls. A buckrake can be mounted in this position. The manure fork available for attachment to the loader arms usually holds 200-250 kg (4-5 cwt approx.) of manure although the loader itself is generally capable of lifting a much heavier weight than this.

The manure spreaders used are of two main types, those that are wheel driven and those that are power driven. In the former, all the mechanisms of the machine are driven by land wheels, whilst the latter are driven by the p.t.o. shaft from the tractor. The power drive machine has an advantage over the other when field conditions are bad because it is not affected by wheel slip.

What is required of a manure spreader is that it should hold a reasonable quantity of manure for transporting and that it should spread this manure evenly across the field at different rates of application as desired. These requirements are generally met with the types of spreaders available; and apart from the way in which they are driven, probably the main difference between most makes is in the amount of manure that can be carried. There are exceptions which will be mentioned later.

The quantity of manure carried is usually not less than 1·5 tonnes in the smallest spreaders, whilst the largest carries about 4 tonnes. Generally, the rates of application of the manure applied to the field can be varied from between 12 and 60 tonnes per hectare (5-25 tons per acre) in 5-tonne increases.

The Spreader Mechanisms

The spreader is strongly constructed because the work it carries out is heavy, but essentially it is a low trailer having a moving conveyor on the floor and shredding and spreading devices at the rear.

The conveyor consists of angle iron slats positioned crosswise on the spreader floor and the ends of these slats are fixed to endless chains, see Fig. 116. If this conveyor can be made to move rearward, any manure that is lying on the floor will also be carried rearwards and towards the spreading mechanisms.

180

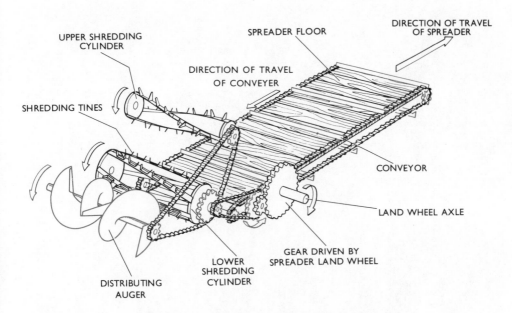

Fig. 116. The drive, feed, shredding and spreading mechanisms of a
wheel-driven manure spreader.

The spreading mechanisms are driven by one of the land wheels through gears and chain drives. They rotate at high speed, and as the manure is fed rearward, the shredding cylinders, which are fitted with tines, pull out and chop up the manure. The manure is thrown on to the distributing auger which spreads it widely on the field. The width at which the manure is spread depends to a large extent on the width of the spreading mechanisms and their design and speed, but it is usually between 2·5 and 6 metres (8 and 20 ft approx.).

The mechanism that moves the conveyor rearward and regulates the quantity applied to the land is shown in Fig. 117. This is driven by the other land wheel and operates as follows:

(a) The land wheel rotates and as it does so the cam wheel also rotates.

(b) The cam wheel is made with three cams: as each one comes against the cam roller the roller is pushed forward.

(c) This causes the ratchet drive arms to move forward.

(d) These arms are attached to plates which hinge in the shaft carrying the ratchet wheel. In between these plates is a ratchet pawl which is in engagement with the teeth of the ratchet wheel. Movement of the ratchet drive arms therefore also causes movement of the ratchet wheel.

(e) The conveyor, being carried on sprockets attached to the same shaft as the ratchet wheel, must also move and its movement is rearward.

(f) Movement ceases when the cam roller rides over the cam on the cam wheel, and the return spring draws the drive arms rearward so that the ratchet pawl takes up another position on the ratchet wheel. This then gives intermittent movement to the conveyor which is taking the manure towards the spreading mechanism.

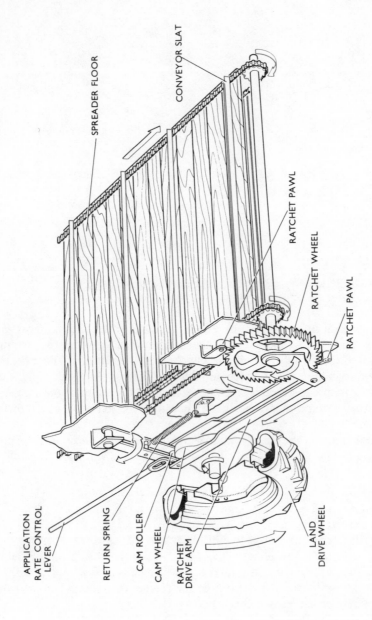

SPREADER FLOOR

CONVEYOR SLAT

RATCHET PAWL

RATCHET WHEEL

RATCHET PAWL

APPLICATION
RATE CONTROL
LEVER

RETURN SPRING

CAM ROLLER

CAM WHEEL

RATCHET
DRIVE ARM

LAND
DRIVE WHEEL

Fig. 117. The feed regulating mechanism of a wheel-driven manure spreader.

Rate of Application of Manure

It is necessary to be able to vary the quantity of manure applied to a field. How much is applied to a field depends on how much a farmer may have available and how much he thinks is required. From the point of view of the machine doing the job, it is desirable that it should be able to apply different rates if required.

A method of varying this rate of application is by controlling the position of the cam roller in relation to the cam wheel. Figure 117 shows an application rate lever which is attached to the arms carrying the cam roller. This lever can be set in a fixed position by the operator. Because the lever is slotted at the end shown, the forward movement of the cam roller is not restricted, but its return position can be restricted by the end of the slot. We can therefore hold the cam roller in such a position that it can be given more or less movement as required. For example, if the roller is positioned so that it lies at the bottom of the cam, then when the cam wheel rotates the roller will receive its greatest amount of movement. This will cause more movement to the ratchet wheel and likewise the conveyor. If, at the other extreme, the roller is held near to the top of the cam, it will be given little movement when the cam wheel rotates.

By altering the cam roller position, the distance which the conveyor moves during each move it makes is altered. By doing this, the amount of manure fed to the spreading mechanism is varied.

To enable the spreader mechanisms to be put in and out of gear as required, a clutch mechanism is used and this is usually built into one of the drive-wheels.

There is a power-driven type which spreads the manure to the front of the spreader instead of the rear. There is now also a trend towards the use of flails to spread the manure instead of the much-used shredding cylinders and augers shown in Fig. 116.

Many manufacturers are producing spreading attachments that can be fitted to farm trailers and this provides another use for the trailer.

Rotary Manure Spreaders

These machines have been widely adopted due to their simplicity of design and ability to handle a wide range of materials right through from a wet slurry to dry farmyard manure.

The material is stored in a cylindrical tank through which a p.t.o. driven shaft runs supporting chains (Fig. 118). During the filling operation it is advisable to periodically rotate the shaft ensuring that the spreading chain wraps around the main rotor. Once

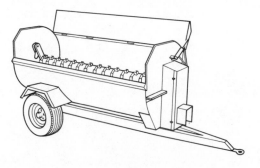

Fig. 118. Rotary manure spreader.

spreading commences the centre shaft rotates at high speed, causing the chains to be thrown out by centrifugal force and bite into the manure, causing it to fly out sideways giving an approximate 10-m bout width. Unfortunately, as the machine empties the rate of application increases and therefore adjustments to the tractor speed are necessary to ensure even distribution.

Routine Maintenance to Farmyard Manure Spreaders

Maintenance to a farmyard manure spreader is no less important than it is to any other farm machine. It is subject to the same type of corrosive action as is a fertilizer distributor. This is due to the acids in the manure which attack the metal parts of the machine. Again it is important to thoroughly clean it when it is not to be used for a long period.

There are greasing points and oil points as on other machines that require daily attention. On power-driven manure spreaders there is usually a gear housing in which gears run in oil and this should be kept filled to the correct level with a suitable grade of oil. It will normally require draining and refilling with new oil annually.

Manure spreaders usually have a number of chain drives that have to be kept properly tensioned to prevent them jumping off their sprockets. These chains and sprockets will require correct lubrication.

It is difficult to clean out a manure spreader at the end of each day's work and probably also unnecessary, but this task should not be neglected, when the machine is not to be used for some weeks or months. All manure should be removed from the spreading mechanisms and other parts of the machine, particularly the conveyor slats and chain. If this is not done, the chain links are likely to corrode together and when the machine is used again they will not ride around the sprockets and will probably snap.

When the spreader has been thoroughly cleaned, all the greasing and oil points should be lubricated and all the metal parts saturated in oil. Old engine or gear oil is suitable for this purpose.

CHAPTER 19

Ground Crop Sprayers

GROUND crop sprayers are used for spraying chemicals to control various types of crop pests and diseases and to kill weeds. These chemicals are usually mixed with water; the resulting materials being one of three types:

Solutions. The concentrate may be supplied as a powder or as a fluid, all the chemical will dissolve to produce a solution which when once formed requires no further agitation.

Emulsions. The concentrate is supplied as a liquid which breaks into small droplets and disperses through the water but it does not actually dissolve. Emulsions are cloudy due to the reflected light of these droplets. Most emulsions are stable and require no agitation after the initial mixing.

Suspensions. The concentrate is supplied as a powder and when mixed with water the small particles do not dissolve but float about within the liquid, therefore constant agitation is necessary to prevent the chemical from settling out.

The Sprayer Circuit

The sprayer circuit is built of the following main components.

Tank. This may be constructed of galvanized steel or plastic so as to prevent corrosion caused by the chemicals used.

Pump. This is usually of the roller vane type (Fig. 119) or of the diaphragm type (Fig. 120) and can deliver liquids up to the rate of approximately 100 l per min at pressures of 4 bars. Although one does not require such outputs for spraying the excessive delivery is necessary to ensure adequate agitation by supplying excess liquid

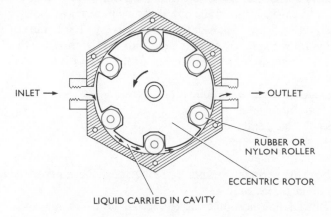

INLET ⟶

⟶ OUTLET

RUBBER OR
NYLON ROLLER

ECCENTRIC ROTOR

LIQUID CARRIED IN CAVITY

Fig. 119. The action of a roller vane pump.

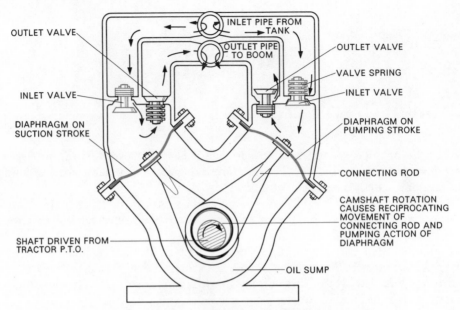

Fig. 120. A sectional view showing the principle of operation of a
diaphragm pump.

back to the tank. An adjustable spring-loaded pressure-relief valve is incorporated in
the pressure side of the circuit to control spray pressure.

Control valve. This controls the flow of liquid in three ways. There is an "on"
position which allows liquid to be drawn from the tank to the pump, which delivers
the material via the pressure-relief valve to the booms to achieve spray. On larger
sprayers (8 m plus), the supply to the booms may be divided into two or three pipes,
each one having its own tap. This provides the means to utilize part or complete bout
widths.

An "off" position stops the delivery of liquid to the booms but to prevent dribbling
from the nozzles the suction side of the pump is partially connected to the boom supply.
This creates a "suck-back" from the booms which in turn avoids dribbling.

Finally a "mix" position enables the total delivery (100 l/min) of the pump to be
diverted back into the sprayer, so that chemicals can be mixed with the water, also in
this position it is safe to leave the sprayer without the p.t.o. running as no syphon
effect can occur.

Filters are usually fitted in the following places.
(a) at the tank filling hole,
(b) between the tank and pump,
(c) in the jet assembly.

Power fill. This permits the sprayer to be filled from private water sources, i.e.
reservoirs, storage tanks, etc.

A detachable suction hose may be connected to a junction on the suction side of
the pump so that when the p.t.o. is engaged, water is drawn from the water source

into the sprayer tank. This can result in a considerable time saving during the spraying operation.

A typical spray circuit is shown in Fig. 121a in the spraying position. Liquid is drawn from the tank through the suction filter; after passing through the pump the

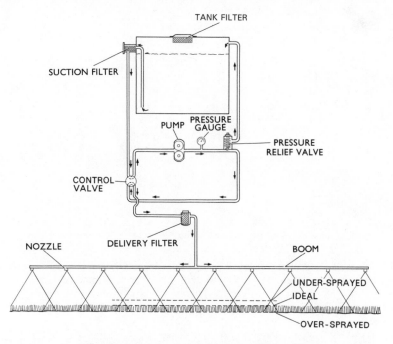

121a. A sprayer circuit showing the spraying position.

liquid flows through the control valve and delivery filter to the boom. A pressure gauge is fitted on the delivery side of the pump in order that the spraying pressure can be accurately set. The pressure-relief valve can be adjusted to set the spraying pressure, and because the pump is always pumping at a greater pressure than is required, the relief valve opens to control the pressure and allows the excess liquid to recirculate back to the tank. This liquid return is used for agitation of suspensions to prevent any solids from settling. Figure 121b shows the circuit in the "suck-back" position, when the pump sucks back liquid from the boom and returns it to the tank.

Rate of Application of Liquid

Application rate is controlled by forward speed, liquid pressure and nozzle size.

This means that by varying any one of these factors the application rate will be affected. Because there are three factors that can be varied either individually or together, it is possible to have a wide range of different application rates. However, not all these variations will result in satisfactory application of the chemical in practice. In order to achieve satisfactory application it is usual practice to use the minimum pressure which will give proper droplet formation. Too low a pressure will cause inadequate "atomization" and too high a pressure will cause a high proportion of very

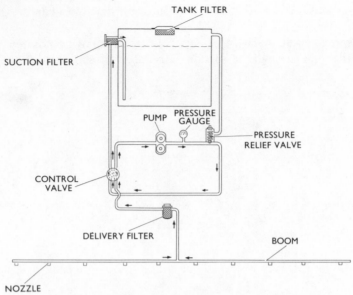

Fig. 121b. A sprayer circuit showing the "suck-back" position.

small droplets, which leads to spray drift. The suitable pressure, together with correct nozzle size and correct forward speed, will give the correct application.

Boom Height

Boom height is determined by the angle of the fan or cone of spray and should be such that an even distribution of chemical is achieved. Normally the distance is approximately 500 mm (21 in.) but this measurement should be obtained from the manufacturer's handbook. It is important to note that the height of the boom is measured in the field, from the top of the crop to the underside of the nozzle. Figure 121a shows the result of incorrect setting.

Preparing the Crop Sprayer for Work

Before the crop sprayer is put to work at the beginning of a season it is essential that it is checked over to ensure that it is in good working order and that it will accurately apply whatever quantity of spray chemical it is required to. The importance of this cannot be over-emphasized because the successful action of most chemicals in use depends on accurate application. Applications in excess of requirements can be wasteful in time and materials.

Carry out the following procedure before using the sprayer.

(a) Check that the pump will rotate freely before attaching it to the tractor.
(b) Check that the relief valve is in order.
(c) Take a look at hoses and joints. Rubber hoses may have perished.
(d) See that the tank is clean and completely empty.
(e) Check that the filters are in good condition. Cracked or otherwise damaged filters may allow dirt through to choke up nozzles.

(f) With the sprayer correctly attached to the tractor, the pump correctly mounted to the power-take-off shaft and properly secured to prevent rotation, the sprayer tank can be filled with clean water.

(g) Open the pressure-relief valve, start the pump drive slowly and any leaks will become evident and should, of course, be rectified.

(h) Check that the pressure gauge is operating satisfactorily by adjusting the relief valve. In some cases there is a gauze-and-wad filter positioned in the pipeline to the pressure gauge. This may require replacement. Examine inside the tank to ensure that liquid is returning to provide agitation.

(i) Remove the nozzles and slowly switch to spray and allow about half the tank of water to wash through the machine. This is important with a new machine as well as an old one.

(j) Replace the nozzles, switch on slowly again and observe the pattern produced by the nozzles. This should not be malformed in any way. Any faulty ones should be dealt with by either cleaning or replacing. If the boom is fitted with fan-type nozzles it is important that the flat spray pattern produced is parallel to the boom. Any that are not must be set correctly by rotating the nozzle tip as required.

Calibration

Having thoroughly checked the sprayer to see that it is mechanically sound and operating satisfactorily it now remains to check to see if it will apply the amounts of liquid per acre that it is supposed to apply. To do this we must "calibrate" the sprayer, and this is an important and essential part of crop-sprayer use.

Firstly, it must be understood that the output of any sprayer remains constant for a given nozzle size and pressure setting. The output of the sprayer can be varied by:

(a) changing nozzle size,

(b) altering the working pressure,

(c) varying the foward speed.

Pressures determine the degree of atomization of the liquid, doubling the pressure will *not* double the output. It will increase output to some extent and will certainly create more drift.

Varying the forward speed will have a direct affect on the output. Application rate is inversely proportional to forward speed, in other words, if the forward speed is halved, the application rate will be doubled.

Sprayer manufacturers provide charts giving details of how to vary application rates. If correct nozzles are fitted and pressures are correct, application rates will vary if forward speed is varied. Therefore, it is important to maintain the correct speed. A speedometer is an obvious necessity on a tractor used for spraying but it must be accurate and can be checked. An important point to mention here is that if the tractor is fitted with narrow wheels for spraying work, and they are a different diameter to its original wheels, the speedometer reading will be incorrect.

To check the forward speed of a tractor, measure the distance in metres covered in 1 minute and divide by 16·6. This gives the speed in kilometres per hour. (If the distance measured in feet is divided by 88 this will give the speed in m.p.h.)

For example, if the tractor travels 50 metres in 1 minute the speed of the tractor will be 50/16·6 = 3 km.p.h. or if the tractor travels 352 ft in 1 minute the speed of the tractor will be 352/88 = 4 m.p.h.

Carry out this check by setting the tractor throttle to position the speedometer pointer at the speed to be checked, say 4 km.p.h., then measure the distance that the tractor travels in 1 minute and so on as detailed above. If it is found that with the speedometer so set your calculation shows that the true speed of the tractor is in fact only 3·5 km.p.h. then you will have to make the necessary throttle adjustment and recheck until you establish the forward speed you require for spraying.

Having established a throttle setting to give a desired forward speed, the crop sprayer can now be calibrated to find out if it will apply the correct amount of liquid per acre. There are a number of methods of doing this. Two will be dealt with here.

(a) *Field Method*

Proceed as follows:

Decide which application rate you wish to check by reference to the manufacturer's chart. For example, if the spray chemical you are to use has to be applied at 225 litres per hectare (20 gal/acre), reference to the manufacturer's application chart will reveal three factors.

(a) Nozzle size   e.g. No. 10.

(b) pressure   e.g. 3 bars (44 lb/in²).

(c) Forward speed   e.g. 6 km.p.h. (4 m.p.h.).

Fill the sprayer tank to a known level, say 100 l (22 gal). Set the pressure at 3 bars (44 lb/in.²) and travelling at 6 km.p.h. (4 m.p.h.) spray an area equal to 1/10 hectare. The distance to be travelled to spray this area will be equal to:

$$\frac{1000 \text{ m}^2}{\text{effective spraying width}}$$

If, for example, the effective spraying width is 10m (32.8ft) then the travelling distance to cover the area will be $\frac{1000}{10} = 100$ metres.

Note: 1 hectare = 10,000 m².

Effective spraying width = number of nozzles × distance apart.

Having sprayed the known area the sprayer tank can be refilled to the initial level measuring the quantity used. As the test was conducted over 1/10 ha, multiplying your result by 10 will give the application rate per hectare.

Variation from the desired application rate may be compensated for by adjusting forward speed, i.e. a 10% excess application can be corrected by travelling 10% faster (6·6 km.p.h. instead of 6 km.p.h.).

The above calibration method provides an overall check on the output of the crop sprayer but does not account for any differences of application rate between individual nozzles in the boom. Nozzles do wear, and some may wear more than others. It is sensible to check them periodically and this can be done simply as follows:

(b) *Static Method*

With a quantity of clean water in the sprayer tank, put the pump in operation, set the working pressure correctly, switch to spray and with a small container, say half a litre or half-pint size, time how long it takes each nozzle to fill the container. By doing this it will be readily seen whether any nozzle is applying quantities of liquid greatly in excess or below that of the others. All nozzles should fill the container in the same time if they are all in equally good condition.

Routine Maintenance to Crop Sprayers

Generally, there are few points on crop sprayers that require lubrication attention. The sprayer pump may be fitted with one or more greasing points that require daily attention when the sprayer is in use. It is important not to overgrease these points otherwise there is a likelihood of grease finding its way into the liquid circuit and blocking filters.

Throughout a working day a sprayer operator is likely to have to remove nozzles and filters for cleaning. These should be rinsed in clean water if possible and not wiped with rags or poked with pieces of wire. When removing these parts the operator should avoid getting his hands contaminated with the spray liquid (see Chapter 30, Safety on the Farm).

Spraying weeds in farm crops usually calls for the use of different types of spray chemicals because the different weeds cannot all be killed by one type of chemical; furthermore, some chemicals will not only kill weeds but will also kill the crop in which the weeds are growing. Care must therefore be taken when applying these chemicals to make sure that the right crop is sprayed with the right chemical.

When it comes to changing from one type of spray chemical to another, it is important that the spraying machine is thoroughly cleaned so as to remove all traces of the chemical that may be harmful to the next crop to be sprayed.

The following procedure should be adopted when changing from one chemical to another.

(a) Drain off any liquid still in the tank.

(b) Put about 90 litres (20 gal) of clean water in the tank and add 1 kg (2·2 lb) of washing soda per 45 litres (10 gal) of water. This will serve as a cleansing detergent.

(c) Spray this liquid through the nozzles on some waste land.

(d) Put another 90 litres (20 gal) of clean water in the tank, remove a nozzle from each end of the boom and spray on waste land until the tank is empty.

(e) Remove all filters and rinse clean if necessary.

The same procedure should be adopted when the spraying machine has completed its season's work and is to be stored away. This is not only because traces of the chemical may remain in the sprayer but also because some chemicals are corrosive. For end-of-season attention we must add to the list the following:

(f) Wash down the outside of the sprayer.

(g) Remove nozzles and filters and store safely.

(h) Release the pressure on the relief valve.

(j) Ensure that no water is left in the pump.

(k) Saturate the inside of the pump with lubricating oil.

When emptying sprayer tanks of spray chemicals and washings care should be taken not to put it where it is likely to be harmful to humans, livestock, wild life or fish.

Health and Safety (Poisonous Substances) Act, 1975

This safety act has been introduced to protect all persons involved with spray chemicals. Materials which are considered dangerous are listed as specified substances which are then subdivided into four parts according to their toxicity. Anyone using spray chemicals MUST familiarize themselves with this act which sets out

1. Minimum age of users.
2. Protective clothing requirement.
3. Limits as to amount of usage (hr/day).
4. Training of operators.
5. Records of chemicals used.
6. Notification of illness.

This law covers all persons whether employer, employee or self-employed and all parties are liable to prosecution should any infringement occur.

CHAPTER 20

Mowers

MANY mowers used on farms are of the cutter-bar type and they may be either mounted directly on and completely carried by the tractor, or mounted directly on to the tractor but partially carried by castor wheels at the rear. The cutting mechanism is driven by the tractor power take-off shaft except on older wheel-driven mowers.

Cutter-bar Action

Figure 122 shows part of a cutter-bar. The knife, driven by a *pitman* or connecting rod, lies through the *fingers,* which are bolted to the cutter-bar back. The *knife sections* slide to and fro over the flat faces of the fingers, which are known as *ledger plates,* from the centre of one finger to the centre of the next, the length of travel being 7·63 cm (3 in.). The action is similar to a pair of scissors with one blade kept stationary and the other moving. As the knife sections slide over the ledger plates the blades of grass are sliced off. *Thumbs* (or "keeps") are fitted to prevent the knife from "riding up" during the cutting action and slotted *wear plates* are built into the assembly to take the backwards thrust on the knife.

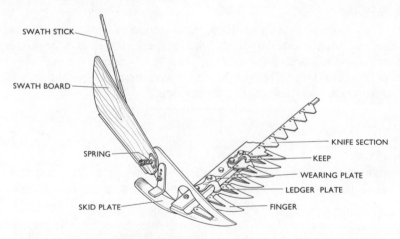

Fig. 122. Cutting mechanism of a mower cutter-bar.

Sharpening and Adjustments for Wear

1. *Knife*

The knife sections become blunt during use and must be resharpened. This can be done with a second cut file or a file known as a reaper file which is specially made for this purpose.

When the blade has been resharpened repeatedly it assumes a triangular shape and it is then usually rejected. Figures 123a and 123b show the removal of the old blade. In the first instance the blade is of the "proud back" type: the blade is a loose fit in the vice and the sharp blow behind the rivet provides a shearing action. In the second instance a "flush back" blade is shown. The knife is rigidly held in the vice and a blunt or square-ended sett is used to apply the force necessary to shear the rivets. Re-riveting is a straightforward task and provided that the rivet is a snug fit in the knife and section and a good hard blow is initially used to swell the rivet, a satisfactory job can be done without a rivet sett, although the latter does make a neater job.

2. *Thumbs or Keeps*

Thumbs or keeps which are usually malleable should be adjusted down till they almost touch the knife. Shims are sometimes fitted under the retaining bolts and removal of one or more of these will take up any "play". If shims are not fitted, the thumb (or keep) is bent downwards by a light tap with a hammer.

3. *Wear Plates*

Wear plate retaining bolt holes are usually slotted to allow fore and aft adjustment. A slight amount of play should be allowed to prevent seizure. It is advisable to use all the mower knives in rotation rather than to keep one as a spare. Since the thumbs and wear plates will have been adjusted as the knives have become worn, any new knife kept as a spare will not fit into the cutter-bar when any breakage calls for a replacement.

4. *Cutter-bar Lead*

A cutter-bar should be at right angles to the direction of travel with the machine in work; because of this it is necessary to give the outer end of the bar a forward lead over the inner end when the machine is stationary. This lead is usually 6 mm (¼ in. approx.) per 30 cm (1 ft approx.) length of bar. When the machine is in work the pressure of the crop on the bar brings it back to the correct position.

Working or Field Adjustments

1. *Height of Cut*

Fully mounted type. The height of cut is controlled by limiting the drop of the mower relative to the tractor; usually by a chain fitted to the linkage mechanism.

2. *Cutter-bar Tilt*

A tilt adjustment is provided for the cutter-bar; in normal use the fingers should slope slightly downwards but it may be found necessary in matted growth to tilt them slightly upwards.

3. *Relationship of Forward Speed to Cutting Mechanism Speed*

This is a most important factor in cutting efficiency. Since a wide variety of crop conditions will be met, only general rules can be given. The cutting mechanism speed

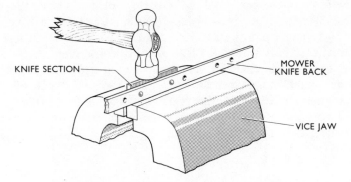

Fig. 123a. Removal of a "proud back" blade.

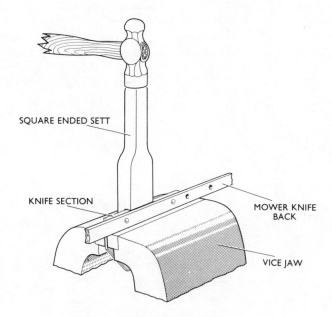

Fig. 123b. Removal of a "flush back" blade.

should be kept as low as practicable (by the use of a low engine speed) and the forward speed should be kept as high as practicable by the use of a high gear, but in difficult conditions the cutting mechanism speed should be increased by increasing the engine speed and the forward speed reduced by the selection of a lower gear.

Machine Safety Devices

The cutter-bar is usually protected from damage by a "break-away". This may take the form of a spring-loaded release mechanism or a shear bolt. If the cutter-bar strikes an obstruction, the safety device releases and the cutter-bar swings back almost into line with the tractor. Many of these devices are self-latching and it is only necessary for the tractor driver to reverse in order to reset the breakaway, but, of course, the mower must be lifted out of work and the obstacle avoided. The cutting mechanism is protected from damage by an over-rider or "clatter" clutch in the drive. If a stone, stake or other obstruction jams in the knife and finger assembly, the line of drive is broken and the noise of the slipping clutch alerts the tractor driver.

The pressure on spring-loaded clutches should be released during storage periods. If they are of the metal to metal type, the mating faces should be lightly oiled to prevent rusting. The minimum practicable spring pressures should be used in the field.

Slow rates of work and susceptibility to blockage with cutter-bar mowers has led to a considerable change to alternative types of mowers.

Disc Mowers

Fully mounted on the tractor's three-point linkage and driven by the p.t.o. these machines consist of four small diameter discs rotating at very high speed (3500 r.p.m.). Attached to each disc are two small blades which may be reversed or replaced when worn. The drive from the p.t.o. is usually transferred by "V" belt down to the cutter-bar where gears, running in a sealed casing, transmit the power to the individual discs, two of which rotate clockwise and the outer pair anti-clockwise (Fig. 124).

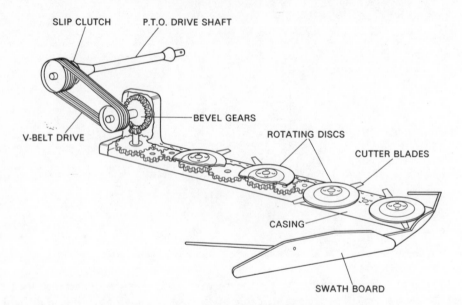

Fig. 124. The drive arrangement on a disc mower.

The usual safety devices such as slip clutches and spring-loaded release mechanisms are also provided on these mowers.

Drum Mowers

Similar in operation to the disc mower but requiring considerably more power to drive them are drum mowers. These, however, consist of two large diameter drums rotating at a lower speed, the drive to these being from belts or shaft and gears running across the top of the machine. A saucer underneath each drum runs on the ground (Fig. 125) and the clearance between drum and saucer may be varied to alter cutting height. Again, safety devices are provided on these machines for protection against damage.

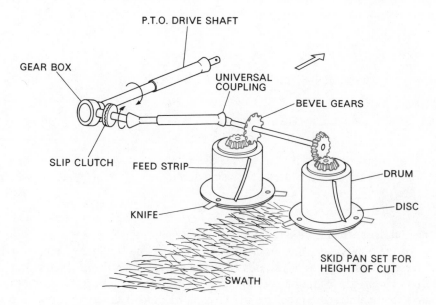

Fig. 125. A drive arrangement on a rotary drum mower.

Flail Mowers (Fig. 126)

These machines are trailed on off-set hitches behind the tractor and consist of a horizontal rotor on to which are mounted four banks of free swinging flails which are held at right angles to the rotor by centrifugal force but may fold back should an obstruction be struck. Hay or silage blades (Fig. 127) may be fitted dependent on circumstances.

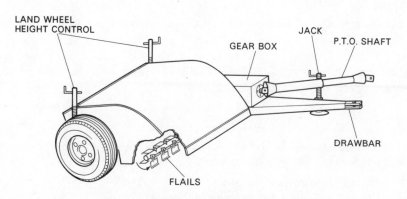

Fig. 126. A flail mower

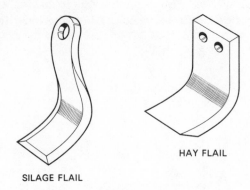

HAY FLAIL

SILAGE FLAIL

Fig. 127.

Height of cut is controlled by varying the pivot point of the hydraulic ram connected to the axle, this ram in turn is used for raising and lowering the machine in and out of work.

Routine Maintenance of Mowers

Some of the points regarding maintenance to mowers have already been mentioned and if these points, such as adjustment to wearing plates and keeps, maintenance of the correct cutter-bar lead, and resharpening and resectioning of knives are attended to, many other maintenance problems will be avoided. Incorrect setting and adjustments to these are the cause of knife and connecting-rod breakages and bad cutting.

In average cutting conditions, knives require replacing about every 2 hours. Other maintenance during work will be confined to lubrication and perhaps adjustment to drive belts where fitted and slip clutches.

On all mowers there are greasing and oiling points that require daily attention, but it is usually recommended that knives are lubricated in their guides with oil every 2 hours, connecting rod ends also usually require 2-hourly lubrication.

Other Types

As disc, drum and flail mowers all rely on centrifugal force to hold the blades in work, high-speed rotation of these machines is such that balance is critical. A broken or missing blade will result in severe vibration which if allowed to persist may cause serious damage. Always replace blades as a pair, fitting one blade directly opposite to the other. In this way balance is maintained.

General

Suitable attention should be given to the mowing machine when the cutting season is finished and before it is put into store. The following is recommended:

(a) Thoroughly clean down the mower.
(b) Examine the mower and if any parts require replacing it is wise to do so at this stage if time permits.

(c) Knives should be cleaned and repaired, coated with anti-rust preparation and stored safely away.

(d) If a belt-driven mower, slacken off belt tensions.

(e) Release the pressure on slip clutches and apply oil to the metal faces.

(f) Apply anti-rust preparation to the cutter-bar assembly and other metal parts if required.

(g) Store in a dry shed.

CHAPTER 21

Haymaking Machinery

AFTER a crop of grass has been cut for making into hay, the immediate problem is to get it into a condition which will allow it to be safely stored, either in a stack or under building cover. This means that most of the moisture in the grass must be removed before this can be done, otherwise when stored in bulk the hay will go "wrong". It will either heat up and may fire or it will go mouldy and rotten.

When grass is cut its moisture content is about 75% but to store it safely this moisture must be reduced to around 16-20%. At the same time as much of the food value as possible must be retained in the hay because excessive weathering, that is, being exposed for long periods to sun and rain, leaches out of the hay much of its value. Haymaking should, therefore, take place over as short a period as possible to get the best results.

To enable a crop of grass to dry out quickly the swath left by the mowing machine must be shaken out to allow the sun and wind to act on as much of the grass as possible. a mower swath that is unmoved will lie thickly on the field surface and dry only on the top.

To assist the field operations of shaking out and moving the swath a variety of machines are used. These machines are *tedders, turners, side-delivery rakes, rollers* and *crimpers*.

The Tedder

These machines are often used immediately or soon after the crop has been cut and their action is such that they leave the swath in a loose fluffed-up condition thus allowing the air to pass through. Tedders may be either wheel driven or power driven and may be designed to move one or two swaths at one time. The action of the tines is shown in Fig. 128 and there is usually an adjustment whereby the angle of the tines can

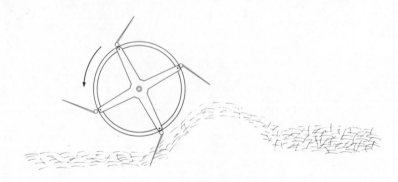

Fig. 128. Tedding action.

200

be altered. This alteration affects the throw of the crop. There is also an adjustment which allows for setting the height of the tines in relation to the ground.

Swath Turners

These are the machines used for turning the swath completely over so that the underside is exposed. They are generally designed to do two distinct operations: swath turning, and by making a simple alteration to the machine, side-delivery raking, when two swaths are put into one. This latter job usually takes place prior to baling or sweeping. A machine capable of doing the two operations is known as a combined swath turner and side-delivery rake.

Two types of these machines are in general use, one of them making use of what is known as *rake bars* which move the swath, whilst the other uses *finger wheels*. Figure 129a shows the action of the rake bar when being used for turning swaths, whilst Fig. 129b shows the arrangement when side-delivery raking is taking place. A centre

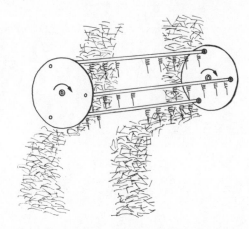

Fig. 129a. Arrangement of rake bars when turning swaths.

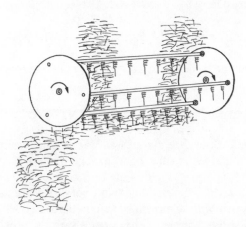

Fig. 129b. Arrangement of rake bars when side-delivery raking.

portion of the rake bar is removed when swath turning is to be done. This type of machine is generally wheel driven.

The finger-wheel machine is of a different construction entirely and its method of driving is also different although it can perform the same two operations. There is, however, more tendency for this machine to roll the swath together, although this is not of any serious disadvantage. The machine is made up of a number of spring-tined wheels, usually four or six, which are mounted in a framework and in such a position that they are at an angle to the line of travel. Each wheel is independently mounted and is caused to rotate by the tines striking the field surface. The faster the machine is towed the faster the wheels rotate. The machine is therefore capable of high-speed operation up to 13 km.p.h. (8 m.p.h. approx.).

When the different operations are to be carried out, the finger wheels are arranged either in two groups or in a single line and the angle of the machine to the line of travel set to cover the required width for contacting the swaths. Figure 130a shows the

Fig. 130a. Finger wheels positioned for swath turning.

arrangement for swath turning whilst Fig. 130b shows the position when side-delivery raking.

These machines may be trailed or mounted and may be mounted to either the front or rear of the tractor.

Roller Crushers

If the stems and leaf of the crop are bruised, sap is more easily and quickly removed by the action of the air and sun, hence the whole operation of haymaking can be speeded up. The roller crusher is a machine designed to carry out this bruising and it is used immediately the crop is cut. Sometimes two operations are carried out at one time using a tractor equipped with a mowing machine and a roller crusher, and they are arranged so that whilst the mower is cutting one swath the previous cut swath is being bruised.

Fig. 130b. Finger wheels positioned for side-delivery raking.

The action of the roller crusher is to pass the crop between steel or rubber-faced rollers which rotate at high speed and bruise. In some cases the rollers may have cleated or grooved surfaces which provide more effective bruising. Figure 131 shows the action of this type of machine. These machines are best used when the weather is certain to be fine and sunny. Crops that are rained on after being bruised and dried are much more liable to spoil.

Crimpers

These machines are of a similar type to roller crushers but, because of the way in which they leave the grass crimped, the swath is probably left in a better condition for allowing the air to pass through. Figure 132 shows the action of this type of machine.

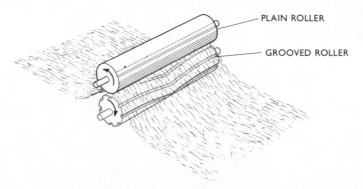

PLAIN ROLLER

GROOVED ROLLER

Fig. 131. The action of a roller crusher.

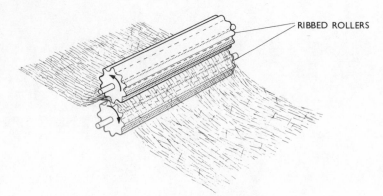

Fig. 132. The action of a crimper.

The Hay Rake

This implement is used for raking hay and putting it into windrows prior to collection or baling, and it may also be used for raking corn stubble fields to clear them of straw. It consists of a framework carrying a large number of curved teeth suspended vertically so that forward motion exerts a combing effect over the ground and cut crop. Tractor rakes are wide implements, usually much wider than the normal farm gateway, so they are constructed so that they can be reduced to a narrow transport width. This is necessary when transporting the rake on main roads.

The modern tractor rake is equipped with a self-lift mechanism operated by the tractor driver. When sufficient hay is collected by the rake tines, the lift is operated and the tines rise and fall leaving behind the collected hay.

Although many tractor rakes are still used, the finger wheel side-delivery rake is quite often used for the same type of job, this being just as effective, if not more so, for cleaning stubbles and hay fields.

Routine Maintenance for Haymaking Machinery

Maintenance to the various types of haymaking machines such as tedders, swath turners, etc., consists mainly of lubrication attention. This will consist of daily or twice-daily greasing or oiling of the various bearings.

CHAPTER 22

Balers

THE pick-up baler is designed to pick up hay or straw from a swath or windrow and compress it into bales which are then tied with twine. Balers produce a rectangular-shaped bale and a typical bale size would be about 35 × 45 × 90 cm (14 × 18 × 36 in. approx.). The length of the bale can be varied and whilst 90 cm (36 in. approx.) would be the longest, the smallest could be about 45 cm (18 in. approx.). The weight of the bales will vary according to their size and density. It is possible on all balers to alter the bale density and this should be set according to the conditions of the crop or other requirements. Bale weight will also be affected by the type of material being baled, but generally the maximum weight of bales produced by this type of machine is in the region of 36 kg (80 lb approx.).

Operation of the Pick-up Baler

The basic operation which produces the bale is essentially the same in all makes. Figure 133 shows a plan view of the main working parts of a typical p.t.o. driven pick-up baler. The arrows indicate the path taken by the crop as it passes through the baler.

When the flywheel rotates, the crankshaft also rotates thus causing the ram to move back and forth. Whilst this takes place, the packer fingers move in and out of the bale chamber. The feed auger rotates continually.

The Pick-up Reel

This is a spring-tined reel which is of sufficient width to take a normal swath of hay or straw. The tines rotate so that they lift the hay on to the *stripper plates,* when it comes into contact with the *feed auger.* It is necessary to be able to adjust the height of the pick-up reel so that it may operate satisfactorily in different field conditions. A control is therefore provided to make this adjustment and the reel tines should be set so that they pick up the crop cleanly and do not strike the ground. If they are allowed to strike the ground, they may be damaged and there is also risk of stones, clods and turf being taken into the baler with the crop.

To prevent any heavy object being taken into the baler and to protect the reel should it strike obstructions in the field, a slip clutch is fitted to the drive shaft. This is often a ratchet plate clutch as shown in Fig. 134 and it is adjusted by varying the spring pressure.

A side view of the pick-up assembly including the feed auger and a windguard is shown in Fig. 135. The crop is left on top of the stripper plates to come in contact with the feed auger when the pick-up tines move below. The windguard covers the full

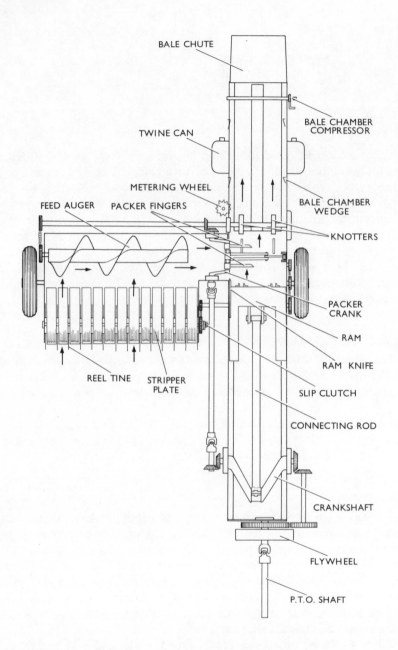

Fig. 133. A view of the layout of the main working parts of a pick-up baler.

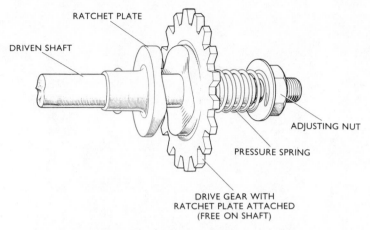

RATCHET PLATE

DRIVEN SHAFT

ADJUSTING NUT

PRESSURE SPRING

DRIVE GEAR WITH
RATCHET PLATE ATTACHED
(FREE ON SHAFT)

Fig. 134. A ratchet plate or "clatter" slip clutch.

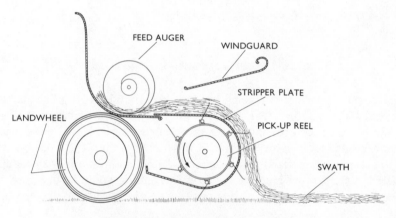

FEED AUGER

WINDGUARD

STRIPPER PLATE

LANDWHEEL

PICK-UP REEL

SWATH

Fig. 135. Side view of the pick-up assembly.

width of the pick-up and is, as its name suggests, for use in windy weather to prevent the crop blowing about as it feeds into the auger.

The Feed Auger

This auger rotates continually and moves the crop towards the bale chamber. It is arranged so that it can float, in other words, rise and fall, and this is done so that it may adjust its position according to the volume of crop fed into it.

The auger may also be fitted with a slip clutch to protect it against overload.

The Packer Arms

It is the *packer arms* that pull the crop into the bale chamber where it comes into contact with the ram. Their action is to move down towards the crop coming from the feed auger, pull it into the chamber and rise up and out. Figure 136 shows a typical arrangement.

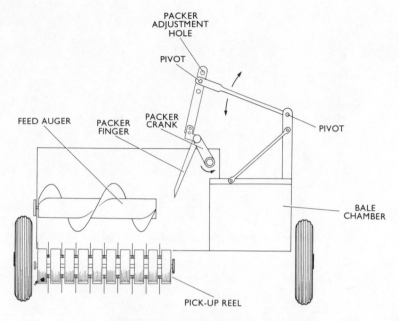

Fig. 136. The position and action of the packer arms.

An adjustment is provided on the packer arms to vary the length of stroke that they make into the bale chamber. This is a necessary adjustment to ensure that the crop material is positioned correctly in the chamber. For example, if the crop is heavy and the packer arms are set to move to their farthest position into the bale chamber, too much of the crop will be packed to the far side. This could result in the production of curved bales. On the other hand, if the crop was light it would be necessary to have the packer arms to move to their farthest position to take it well into the bale chamber.

The Ram

The ram moves forward and backward and its movement is timed in relation to the movement of the packer arms. This allows the ram to be in its backward position whilst the packers are pulling the crop into the bale chamber. When they rise out of the bale chamber, the ram moves forward to compress the material.

The Ram Knife

To sever the crop as it enters the bale chamber, a knife is fitted to the side of the ram and this moves past a shear plate (Fig. 137) fitted to the side of the bale chamber. Every charge of material pulled in by the packer arms is therefore sheared off when the ram moves forward and then pushed along the bale chamber.

The Ram Safety Device

The ram must be protected against all likelihood of damage and this is very often done by means of a shear bolt. This method of protection is used quite a lot on balers

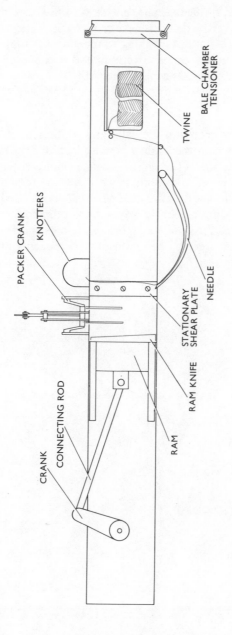

Fig. 137. Side view of the baler.

to protect other mechanisms. It is very simple and makes use of a single bolt or pin, which holds the working part of the machine to the mechanism that is driving it. Should the working part become over-loaded, the bolt or pin or key shears off and the drive ceases to be transmitted. The important thing to remember is that these bolts or pins are designed to shear off under a certain overload and they should never be replaced by any other than those specified by the manufacturer.

The Bale is Formed

Continuous feeding of the crop into the baler and continuous movement of packer arms and ram will result in a continuous flow of material along the bale chamber. This material will eventually pass out of the end of the bale chamber, but before it does we want it tied into bales. To do this, knotting mechanisms are used to tie two bands of twine around each bale. Two knotters are used which are positioned on top of the bale chamber and they operate simultaneously driven by knotter cam wheels. Figure 137 shows the relative position of the knotters on top of the bale chamber and the position of other essential parts. Figure 138 shows various components of a single knotting mechanism.

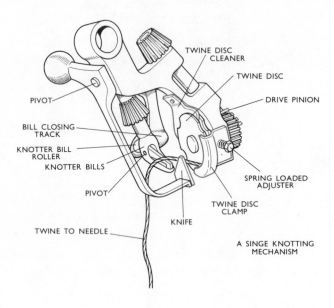

Fig. 138. The various components of a single knotting mechanism.

The twines are held in the knotter retainers and pass over the knotter bills and down through the bale chamber to the needles so that there are two strands of twine situated vertically across the chamber. From the needles, the twine extends to the twine balls held in the twine cans. When the crop is pushed forward by the ram, the twine is also pushed forward and as more twine is needed it is fed from the balls.

Tying Up the Bale

In Fig. 133 a small starlike metering wheel is shown protruding into the bale chamber. This wheel is caused to rotate by the crop passing through the bale chamber and it is connected to a mechanism which trips the knotter and needle drive. When it makes one revolution, the knotter clutch is tripped, the needles rise up through the bale chamber, the knots are tied and the needles fall back to their rest position. The actual tying of the knot takes place by all the mechanisms shown in Fig. 139 carrying out their timed and necessary function. As seen, one end of the twine is held in the twine disc clamp. When the needle rises it brings the other end of the twine up and over the knotter bills and places it in the notch of the twine disc so that there are now two twines lying over the bills and held in the twine disc clamp. At this point all the mechanisms of the knotter have started to rotate or move simultaneously to carry out the tying sequence.

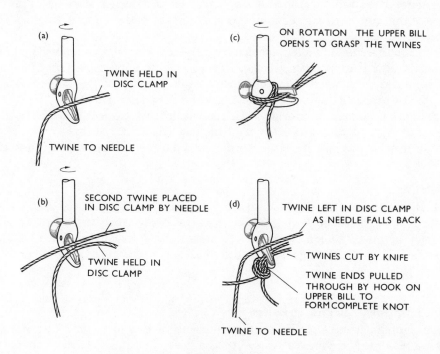

Fig. 139. An illustration of how the knot is formed.

The knotter bills are driven so that they make one revolution and, in doing so, the bills open, grasp the twines then close again when they run on the bill closing track. The next part of the tying sequence is that the twines are cut by the knife, the needles fall back to their rest position beneath the bale chamber and leaves one end of the twine held in the twine disc. The knot is then stripped off the bills by the bale being pushed along the bale chamber. Stripping of the knot off the bills may also be assisted by a stripping device which could have the knife attached to it and thus strips and cuts at the same time.

Figures 139a-d show the action of the knotter bills and illustrates how the knot is formed.

When the bale is formed and tied, it is usually done by timing the mechanisms so that the needles rise to the knotters when the ram is on its backward stroke compressing the hay. The needles rise through slots in the ram and are thus protected from the weight of the crop. A ram stop also rises into the bale chamber at the same time as the needles and, should the needles fail to move down and out of the bale chamber for any reason, the ram will strike the stop on its next return stroke. If this happens, the main drive safety bolt will shear and the drive to the ram will cease. The cause of the needles failing to return must be put right and the safety bolt replaced before the baler is operated again.

Bale Chamber Tension

To increase the density of the bales being produced it is only necessary to increase the resistance offered to the crop as it moves through the bale chamber. This results in more of the crop being packed into a bale so that the bale produced is tighter and heavier—more dense.

Two methods are used to alter bale density. The main one is a compressing device on the end of the bale chamber which can be adjusted to move the sides of the bale chamber towards each other. The closer they are moved together, the more resistance against the crop passing through. This adjustment is shown in Fig. 133.

If the adjustment does not provide sufficient bale density, the density can be further increased by fitting wedges to each side of the bale chamber. As many as three on either side may be used and these offer further resistance to the bales passing through.

The Bale Chute

A short bale chute is usually fitted to the end of the bale chamber and as the bales leave the chamber and slide on to it they can be lifted off by a worker on a bale sledge which may be towed behind the baler. A hitch point is usually provided for this purpose, and if the bale sledge is used, bales can be left in stacked heaps about the field.

Routine Maintenance to Pick-up Balers

The pick-up baler must receive good maintenance if it is to do satisfactory work and last a reasonable number of years. It has a heavy job to do and there are many mechanisms on it that require regular attention.

The lubrication of balers is generally divided into points that require daily, weekly and annual attention. Gearboxes require annual attention, when they should be drained off at the end of the season and refilled to the correct level with a recommended lubricant. Whether or not a mechanism is lubricated daily or weekly will depend largely on the speed at which it rotates or moves. For example, whilst a baler is operated the pick-up reel is rotated continuously and at a reasonable speed, therefore the bearings in which the pick-up reel shaft is carried must be lubricated not less than once a day. On the other hand, the metering wheel shaft rotates very slowly as the hay moves through the bale chamber. Weekly lubrication is sufficient for this.

With some balers it is recommended that the guides on which the ram moves should be lubricated frequently with oil, but for the exact detail of the lubrication requirements of any baler or other machine, the machine operator must refer to the appropriate instruction book.

There are other items of maintenance that a baler operator will have to do depending on how much use the machine gets. It has been explained that during the operation of a baler, as hay is taken into the bale chamber by the packer arms, each charge of hay is sheared off by a knife fitted to the side of the ram (see Fig. 137). This knife becomes blunt through use and must be removed for sharpening. If it is used in a blunt or damaged condition, the hay will not be sheared off cleanly and one side of the bales produced will have a very ragged appearance. Most baler operators carry a spare sharp knife in their tool box and use this as a replacement when required.

The ram is usually fitted with wood runners on which it slides as it moves back and forth in the bale chamber. These runners eventually wear so that the ram becomes a looser fit in the bale chamber with the result that the knife on the side of the ram does not run close enough to the shear plate (see Fig. 133) to give a good cutting action. This will also cause the production of bales with a ragged side.

To compensate for this wear it is usually possible to adjust the runners outwards so that they move nearer to the sides of the bale chamber. Generally, they should be adjusted to such a position that when the knife is alongside the shear plate, there is not more than a clearance of 1 mm (1/32 in. approx.) between their sides.

Other maintenance to be carried out on a baler will include correct tensioning of drive chains and belts. If chains are run with insufficient tension on them, they will tend to jump off the sprocket teeth and this may result in a broken chain. If they are run with excessive tension, the chains will wear and stretch and there will also be an excessive strain on the shaft and bearing. Belts that are given excessive tension will produce the same result. Belts that are run too slack will slip and the shafts that they are driving will not rotate at their correct speed. This information applies to all machines that use chains and belts for drives.

Baler knotters tend to become choked with dirt and rubbish as baling proceeds, and if this is not removed occasionally, knotter troubles can develop caused by trash obstructing the operation of the knotter parts. It is therefore important to occasionally clean off the rubbish.

At the end of the baling season, a baler should be given a thorough clean down and servicing before it is put into storage. The following procedure is recommended:

(a) Remove any bales still in the bale chamber.
(b) Thoroughly clean down the baler removing all hay, straws, seed and other rubbish.
(c) Drain off oil sumps and replenish with new oil.
(d) Slacken the tension on any belts.
(e) Remove all chains for cleaning and oil them thoroughly before replacing them.
(f) Dismantle slip clutches, clean them and reassemble leaving spring tensions slack. If the clutches are the ratchet type, oil the metal faces.
(g) Slacken off the bale chamber tension.
(h) Coat all the bright parts of the baler with grease to prevent rusting.
(j) Lubricate the machine all round.
(k) Store in a dry place. If it has to be kept in an open shed, cover it with a tarpaulin.

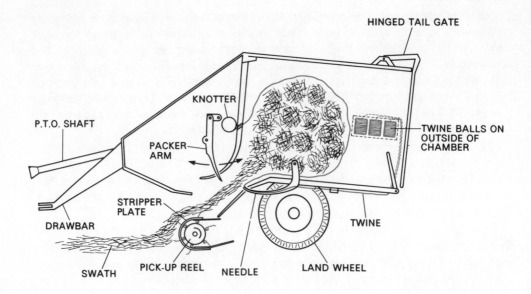

Fig. 140. The press baler.

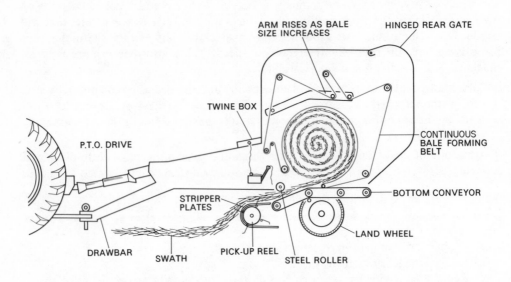

Fig. 141. The roll baler.

Big Balers

Due to the considerable handling involved with the baling operation, manufacturers have produced machines capable of forming bales up to ·5 tonne in weight. Two basic types are available, the press baler which forms a rectangular bale of 2·4 × 1·5 × 1·5 m (2·6 × 1·6 × 1·6 yd), and the Roll baler which forms a cylindrical shaped bale 1·8 m diameter by 1·5 m wide (1·9 × 1·6 yd).

Press Baler

The machine is powered by the tractor p.t.o. and requires approximately 45 kW, somewhat over double that of a conventional baler. Material is picked up by a pick-up reel (Fig. 140) from where it is passed on to a series of packer tines. These force the material into the bale chamber, the front of which reciprocates to compress the bale. Such time as the required density is achieved the forward movement is stopped and three knotters operate to tie the bale which is then pushed through the opened tailgate by the next bale to be formed.

Roll Baler

The advantage of this type of bale is that due to its shape it may be stored out of doors without serious deterioration. Material is lifted from the swath by a pick-up reel and fed into the rolling chamber which consists of an endless belt (Fig. 141) which rolls the material into a cylindrical coil. Tension increases as the bale enlarges and twine is wrapped around the bale to prevent undoing. Finally the bale is discharged by the opening of the bale chamber.

CHAPTER 23

Forage Harvesters

FORAGE harvesters may be categorized into three groups:

1. Single chop machines, which use a flail to cut the crop and produce a chop length of approximately 150 mm (6 in.) and above. These machines may be used for direct cut or picking up wilted material.
2. Double chop machine which, as the name implies, cuts the crop twice, firstly by flail and then again by knife. Again these machines may be used for harvesting direct or wilted material.
3. Precision chop machines which are intended for use with wilted material producing chop lengths from 3 mm to 60 mm (1/8 in. to $2\frac{3}{8}$ in.) approximately.

Single-chop Machines

Figure 142 shows a side view of a forage harvester. L-shaped flails are attached by hinges or links to a rotor which is driven by the tractor p.t.o. The flails fly outwards due to centrifugal force and as the machine is drawn along the flails cut the crop. Due to the draught created the crop is lifted up the delivery chute. A shear plate fitted in front of the flails causes greater or less laceration of the crop depending on its position. In most cases the crop is collected into a trailer and the chute can be adjusted to control

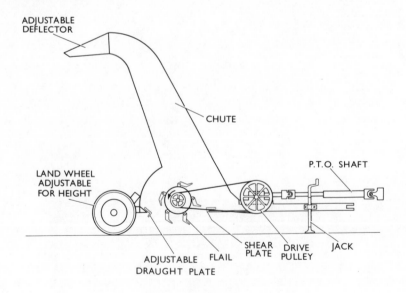

Fig. 142. Side view of a flail-type forage harvester.

delivery and ensure efficient filling. Power to drive the flails is taken from the tractor p.t.o. through a bevel gearbox to a cross-shaft and then by V-belts to the rotor.

Two different types of machine are in common use.

1. *In-line.* The machine is attached directly behind the tractor. Since the rotor is usually narrower than the tractor wheel track some crop is run down by the tractor wheels and it is necessary to make alternate runs in opposite directions. The delivery chute may have vertical adjustment only; in which case the collecting trailer must be drawn directly by the harvester. Automatic or hydraulically operated trailer hitches are available which allow the operator to change a full trailer for an empty one in a reasonable time. If the delivery chute has horizontal as well as vertical adjustment, the collecting trailer can be drawn alongside by a second tractor or attached to the rear of the harvester. The use of a second tractor may be advisable if the output of the machine is very high.

2. *Off-set.* The machine is off-set in relation to the tractor. This allows a field to be cut by the round-and-round method, since there is no treading down of the crop by the tractor wheels.

Working Adjustments

1. Height of cut is controlled by:
   (a) the position of the harvester wheels and/or
   (b) the harvester to tractor drawbar linkage.

Skids are provided at the ends of the rotor chamber to cut down the risk of flails striking the ground when the level wheel(s) drop into a furrow.

2. Laceration is controlled by the position of the shear plate. The smaller the gap between this plate and the flails (when in the working position) the greater will be the degree of laceration. This plate is removed when cutting grass for hay and when using the harvester as a haymaker.

3. Engine speed/forward speed. Since forage harvesters need a lot of power the engine should be run at or near the maximum governed speed. The highest practicable forward speed (i.e. the highest gear) should be used. When the harvester is driven into the crop the latter acts as an air seal but a low forward speed allows air to leak in at the front of the flail chamber.

(N.B. Since not all tractor p.t.o.s run at the same speed, alternative belt pulleys are available to ensure that the rotor speed is correct. The correct rotor speed can be learnt by referring to the maker's manual. Speeds of about 1500 r.p.m. are usual.)

With minor adaptations or attachments forage harvesters can be used for many farm jobs, of which straw chopping, grass cutting for hay, hay turning, bracken chopping and clearing of scrubland in general, cutting kale and maize are a selection.

Double-chop Machines

Figure 143 shows the mechanical layout of a typical double-chop machine. Its method of operation is very similar to that of the single chop, except that once the crop has been cut by the flails it is deposited into a trough where an auger feeds the material sideways to a flywheel on which are mounted usually three blades. These in conjunction with a stationary shear plate cut the grass into shorter lengths (60 mm +) (2½ in. +).

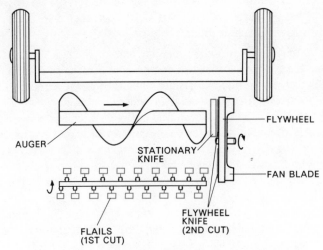

Fig. 143. A view of the layout of a double chop forage harvester.

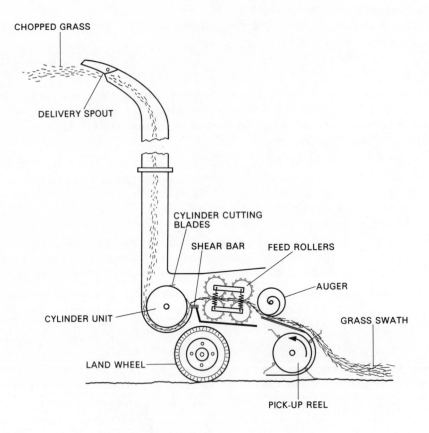

Fig. 144. Precision chop forage harvester.

The flywheel is also fitted with paddles which create the necessary air flow to carry the crop into the trailer.

Working adjustments are also very similar although it is essential to maintain a good knife clearance, say ·3 mm (12 thou. approx.), which is usually obtained by the insertion of shims behind each knife and also correct paddle clearance of about 1 mm (40 thou.) by moving the paddles which are attached by bolts through elongated holes.

Precision Chop Machines

The desire for better quality silage has led to an ever-increasing number of precision chop machines which are capable of providing a short cut material which in turn will give better clamp fermentation.

Wilted material is picked up from the swath by a pick-up reel and gathered in between four feed rollers (Fig. 144) which compress the crop into a tight mass. As the feed rollers rotate the material is presented to a cylinder on to which are mounted a number of blades, usually six, which cut the crop against a stationary shear bar. The air stream created by the blades carries the material from the machine.

Length of cut is controlled by altering the speed of the feed rolls and/or by varying the number of blades in the cutting cylinder. Maintenance of the blade shear-bar clearance and sharpness of the blades is critical.

Sharpening Procedure

The sharpening of blades on a cylinder type machine can be carried out *in situ*. With the p.t.o. running at slow speed a carborundum stone is passed backwards and forwards across the edge of the knives, grinding a sharp edge. The stone is automatically advanced at the end of each stroke.

It is not desirable to have the blades too sharp because the point of the blade would quickly become blunt when in work. Having sharpened the cylinder the shear-bar should be adjusted to give the correct shear bar/cylinder clearance.

Routine Maintenance to Forage Harvester

Lubrication attention is necessary to these machines and will generally consist of daily greasing of the p.t.o. shaft drive, cross-shaft bearings, rotor bearings and wheel bearings. A gearbox is fitted to forage harvesters in which bevel gears run in oil and this should normally be inspected weekly and topped up to the correct level with suitable oil.

The other main points of maintenance on a forage harvester are concerned with the rotor and the V-belt drive. The flails on the rotor should be checked regularly. Any that are broken or missing should be replaced immediately. This is necessary because a complete rotor with flails is accurately balanced and should a flail be lost or broken this balance will be disturbed. The result could be severe vibrations produced on the rotor and damage to the bearings. If a flail has to be replaced for any reason, it must be replaced with one of the same type and only the correct bolts and washers must also be used.

The V-belts which drive the rotor should be checked for wear and alignment. The bevel gearbox mounting is usually adjustable to provide belt adjustment and great care must be taken when retightening the attachment bolts. Due to reaction forces this gearbox can easily become displaced in use: thus the cross-shaft drive becomes out of alignment.

At the end of the season's work the following should be done:

(a) Thoroughly clean the machine.

(b) Inspect the rotor for damaged flails and replace any that require it.

(c) Inspect the rotor bearings for wear and have them replaced if necessary.

(d) Inspect the drive V-belts. If any of these (there are usually three or four) requires replacing, it will be necessary to replace the lot. This is because a single new belt will be a different length to the others which will have stretched during use. The tension on the V-belts should be released.

(e) Drain out the gearbox oil and refill to the correct level with a suitable grade of oil.

(f) Inspect the drive shafts for wear at universal joints and replace if necessary.

(g) Lubricate all greasing points.

(h) Store the harvester under cover if possible.

CHAPTER 24

The Buckrake

THE buckrake is a simple implement used for collecting and transporting grass and green crops for the making of silage. It is also widely used for other farm work, for example transporting farm commodities, in particular bales of hay and straw. Many farmers use it for clearing fields of bales during haymaking and, after baling, the straw left by the combine harvester.

An illustration of a buckrake is shown in Fig. 145. It can be attached to the three-point linkage of the tractor or to the arms of an hydraulically operated fore-end loader. In this way the load collected can be lifted and transported whilst suspended on the lifting arms.

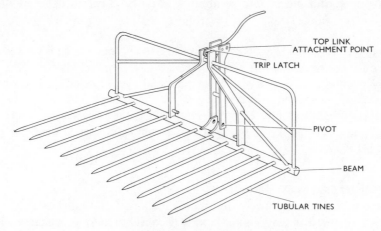

Fig. 145. A tractor buckrake.

Buckrakes are made in various sizes and may be fitted with different types of tines depending on the type of work they are to do, but the standard buckrake is between 240 and 270 cm (8 and 9 ft) wide and has tines spaced 23 cm (9 in. approx.) apart. The tines may be tubular steel or solid spring steel and are about 1·4 m (4 ft 6 in.) long (approx.). This type of buckrake is designed to collect material from two swaths of grass cut by a mowing machine fitted with a 1·5-m (5 ft approx.) cutter-bar.

The field operation of the implement is quite simple; for example, with a rear-mounted buckrake, the tractor is reversed centrally and in line with two swaths of grass, the buckrake is lowered on to the field surface then reversed at a reasonable speed into the swaths. When sufficient grass is collected the rake is lifted and the grass is carried to where it is required. The load is removed by operating a trip mechanism which allows the tines and frame structure to tilt downwards so that the grass slides off when the tractor moves forward. An alternative method of removing a load from a front-mounted buckrake is to use an hydraulically operated push-off device. This is a more suitable unloading method when the material is being loaded into trailers.

Troubles will occur when using the buckrake if it is not properly set and maintained. For example, grass will not move back up the tines if they have been allowed to rust. They should therefore be kept polished. Tines that are bent out of alignment will also hamper the operation by either digging into the ground if they are bent downwards, or causing the grass to roll beneath the rake if they are bent upwards.

When set correctly the tines should slide along the field surface and the beam of the buckrake should be just an inch or two above the field surface. This setting can be achieved by alteration of the length of the top link. Lengthening it will raise the beam and cause the tines to tilt downwards at the points whilst shortening it will lift the tine points and lower beam. Obviously, there is a correct setting to be maintained and this is such that the tines are not tilted sufficiently to cause them to dig in, nor so flat as to cause rolling of the grass beneath them.

Assuming that the buckrake is set correctly, to get satisfactory loading of it, it is also necessary to travel down the swaths in the same direction as the grass was cut by the mower. In this way the grass moves more readily on to the tines. A reasonable working speed is required and this should be about 5 km.p.h. (3 m.p.h.) so that the grass is pushed right back to the frame. Slow working speeds result in small loads being picked up.

With any type of buckrake a certain amount of the crop is left on the field because it falls through the tines, but this need not be excessive. In fact very little is left behind if the implement is used properly and the grass is long. Most difficulty in picking up the grass cleanly is experienced when the grass is short, but buckrakes having the tines spaced closely together can be used for this work. Care is needed when transporting the grass if the travelling has to be done over rough farm roads, otherwise grass may be shaken off the load.

Routine Maintenance to the Buckrake

There are no mechanical drives on a buckrake and so no lubrication points. The maintenance required is very limited and will consist mainly of aligning the tines if they become bent. When a buckrake has finished its work at the end of the season it is worth while to coat the tines with an anti-rust preparation to prevent them from rusting. If this is done, there will be less problems with the crop not picking up properly when the buckrake is next used.

CHAPTER 25

The Combine Harvester

THE combine harvester is used to harvest all types of grain and seed crops and its use has relieved farmers of much of the burden of harvest. It has not taken all the problems out of harvesting, most of which are created by weather conditions, but it has enabled farmers to rescue crops which otherwise may have been lost. Fewer acres of crops are now left to perish in the fields.

The combine was first developed for use in countries where the climate is more suitable for grain production and it was said that the climate in such countries as the British Isles is totally unsuitable for its use. This is not true. Combines will work in quite damp conditions, but where the conditions are not ideal, they must be used in conjunction with a grain drier where the grain can be dried to a safe moisture content for storage.

Its use results in a big reduction of time and labour required to harvest a given crop because one man, or at the most two men only, are needed to operate the machine. Other labour is required to transport the grain to the farmstead, but this is not excessive. There is no stooking, carting, stacking, or any of the other work associated with harvesting when the binder is used. The combine cuts and threshes the grain in one operation.

Combine Types

Three main types of combines are available and these are:
(a) tractor p.t.o. driven combine,
(b) tractor-drawn, engine-driven combine,
(c) self-propelled combine.

The tendency today is towards the use of the self-propelled type because of certain advantages that it has over the other two. It is more or less a self-contained unit propelled and having the threshing mechanisms powered by its own engine. This means that it is not necessary to provide a tractor to drive it and/or pull it. It is usually quite easy to manoeuvre and work in the field because all the controls used to operate it are close to the driver; and when he becomes acquainted with them, he finds it no more difficult to drive than his tractor. Probably the worst hazard confronting the driver of a self-propelled combine is the constant cloud of dust surrounding him whilst he is working the machine. This is much worse when crop conditions are very dry. However, it is possible to have fitted to a combine, an air-conditioned cab which considerably alleviates the dust problem.

Self-propelled combines are made in widths of cut ranging from about 1·5 m to 6 m (5-20 ft), the latter being used on the great grain-growing areas in such countries as Canada. In this country such wide cutter-bars can present problems when it is necessary to move the combine from field to field down narrow country lanes, and 365 cm (12 ft) of cut is a common size.

Of the other two types of combines the main difference is the way in which they are driven but in both cases a tractor is required to haul the combine. In one, the tractor hauls the machine which is driven by an engine, whilst in the other the tractor hauls and drives the combine by the p.t.o. shaft. The width of cut of these types generally does not exceed 2 m (7 ft) but this does not mean that their output is necessarily lower than a larger-cut self-propelled combine. Combine output depends more on the capacity of the threshing and cleaning mechanism within the machine than on width of cut it will take.

Combine Layout

The combine harvester can be described as a mobile threshing machine because it differs little from the stationary thresher except that a crop is fed into a combine much more evenly and continuously as it moves forward. Figure 146 shows the various working parts of a typical self-propelled combine and the flow of the crop through it. Reference should be made to this in conjunction with the following text.

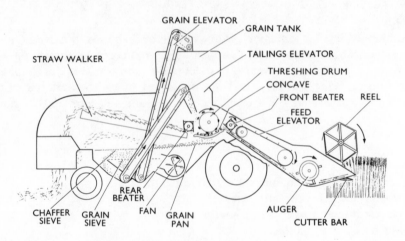

Fig. 146. A section through a combine harvester.

The Cutter-bar

The cutting mechanism on a combine harvester is the same as that on a binder, consisting of a finger bar and reciprocating knife. This method of cutting crops has been used for many years now and, provided that the mechanisms are properly maintained, it is satisfactory. The mower is another machine which employs this method. The knives used on combines and binders operate at a slower speed than the mower knife because the dry straw crop is easier to cut than a swath of grass. Furthermore, it is normal practice to use a serrated edged knife on machines cutting straw crops whilst the mower uses a smooth edged knife. Very often it is found necessary to use a smooth edged knife on a combine if there is a lot of green undergrowth to be cut, but it is better to avoid cutting this undergrowth if at all possible.

Figure 147 shows methods used to provide knife movement. The drive to the crank pulley is usually by a V-belt and this should be kept correctly tensioned to prevent any loss of knife speed.

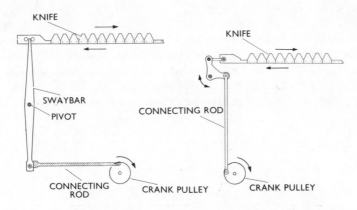

Fig. 147. Methods of providing knife movement.

The height of the cutter-bar is adjustable vertically by mechanical or hydraulic control and it is usually arranged so that the whole cutter-bar bed and reel rises and lowers as a complete unit.

The Reel

The purpose of the reel is twofold. Where a crop is standing well, the reel must steady it against the cutter-bar whilst it is being cut. Where the crop is not standing well, and may be in a tangled and flattened mass, the reel must pull it into the cutter-bar for cutting and feeding into the combine.

Generally, a fairly standard type of reel, known as the pick-up reel, is used and there are few badly laid crops that it cannot effectively pull into the combine, provided that it is correctly set. This type of reel has spring tines fitted to it which can be set at an angle if necessary.

The other type of reel is known as a bat reel and its use is restricted to crops in good standing condition. It is not fitted with tines for pulling in laid crops.

To provide for different crop conditions, certain adjustments can be made to the reel. For example, the reel must not strike the crop any more than necessary otherwise grain will be shed out of the ears and on to the ground. This will happen if the reel speed is too fast. On the other hand, if the reel speed is too slow the crop will be pushed forward by the cutter-bar and fall beneath it. The speed at which the reel rotates can therefore be altered to suit the volume of crop. Alteration is usually brought about by fitting a different size of reel-drive sprocket.

It is also necessary to be able to raise and lower the reel whilst the combine is working and this is done so that it can be held in its correct position according to the height of the crop. It should be just low enough to have a steadying action on the crop. If it is set too low, the straw will be carried around on the reel bars and tines. An hydraulic or mechanical control may be provided to make this adjustment.

One other important adjustment is necessary and this is to be able to move the reel forward and backward if necessary. The usefulness of this is when a crop is laid forward away from the cutter-bar.

The Crop Auger

Transference of the crop into the combine for threshing and cleaning is usually achieved by one of two methods. The retractable tined auger is used as one of these methods together with a chain and slat elevator, whilst the canvas draper-type elevator is used in the other. An alternative method makes use of a series of beaters instead of a chain and slat elevator. An auger and the draper conveyor are shown in Figs. 148a and 148b.

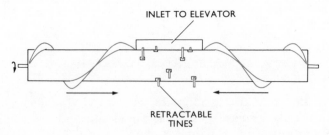

Fig. 148a. A retractable tined auger.

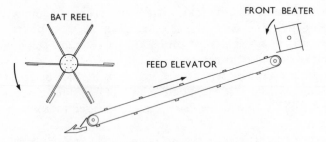

Fig. 148b. A draper conveyor type elevator.

Where the auger is used, as on most self-propelled combines, the spring-tined pick-up reel is also generally used with it and this equipment is perhaps the most suitable for general crop conditions. The auger is a rotating cylinder, webbed on the outside in such a way that the crop is drawn from both ends and towards the centre. Here the crop comes in contact with tines that feed it on to the elevating mechanism. To protect the auger a slip clutch is fitted in the drive mechanism and it should prevent rotation of the auger should any heavy obstruction be met.

The bat-type reel is used with the conveyor-type elevator and, whilst this may not be the best arrangement for combining laid crops, it is probably the best arrangement if damage to grain is to be reduced.

The Front Beater

This is shown in Fig. 146 and its purpose is to feed the crop evenly into the threshing drum. It has an important job to do, therefore it should not be neglected when servicing of the combine takes place.

The Threshing Drum and Concave

From the front beater the grain and straw passes into the threshing mechanism where the grain is beaten out of the ears and separated from the bulk of the straw. This is done by a rotating cylinder fitted with beater bars which rotates above a stationary grid known as a concave. The concave is also fitted with bars throughout its width and it is between these bars and the bars of the cylinder that the grain is beaten out. After being beaten out, the bulk of the grain falls through the concave grid. Figure 149a shows a cylinder used in modern combine harvesters. These cylinders are often referred to as drums. Figure 149b shows an illustration of a concave.

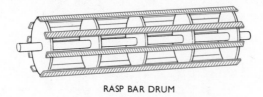

RASP BAR DRUM

Fig. 149a. A rasp bar threshing drum.

CONCAVE

Fig. 149b. A concave.

The type of cylinder shown in Fig. 149a usually has six or eight bars and they are grooved as shown and are called rasp bars. The whole cylinder may be known as a rasp bar drum. In some American machines a peg type of drum and concave may be used. This type results in the straw being much more broken up than is the case in the rasp-bar type.

Drum and Concave Settings

On all combine harvesters there is provision for altering the speed at which the threshing drum rotates. There is also provision for altering the position of the concave in relation to the drum. By altering the drum speed and the position of the concave the extent to which a crop is threshed can be varied. These adjustments are necessary because different crops require different degrees of threshing to remove the grain or seed from the heads. The faster the speed at which the drum rotates the more severe the threshing. The severity of threshing is also increased as the clearance between the drum and concave is reduced. Manufacturers usually state drum speeds in r.p.m. but it is really the *circumferential* or *peripheral* speed that matters. This means that a drum speed given in the case of one machine will be entirely different to that of another make. It will be found that the diameters of the drums are also different. A small-diameter drum will rotate at a higher speed than a large-diameter drum but the circumferential speed may be the same.

Generally, it will be found that crops that produce large-sized grain or seeds are threshed satisfactorily with the slower drum speeds, whilst smaller seeds such as grass seed require a fast drum speed. Figure 150 shows the position of the drum in relation

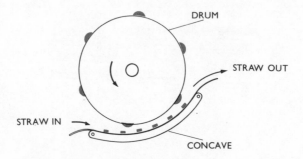

Fig. 150. The relative position of threshing drum and concave.

to the concave. One usual method of altering the drum speed is to fit a different-sized drive sprocket to the drum shaft, but the tendency is now towards the use of variable speed pulleys. This is a very useful and convenient method and its use has also been extended to other drives on the combine. For example, it is also used to provide an infinite number of forward speeds of the combine and likewise it is also used in some cases to provide an infinite number of speeds on the reel. Figure 151 shows an illustration of this device.

It is so made that the pulley sheaves can move along the shaft on which they are keyed, thus the inner sides of the sheaves which grip the V-belt can be moved in or out and be held in any desired position. The arrangement shown in Fig. 151 allows for adjustment of the sheave on the bottom pulley by a hand wheel on a screw thread. The sliding member of the sheave on the top pulley has its position altered against or with the assistance of spring pressure. If the bottom pulley is closed by turning the hand wheel, the belt will move outwards to find a position where its width is satisfactorily gripped. At the same time the belt moves inwards on the top pulley against the spring pressure. The effect is to produce varying speed ratios between the two pulleys, a ratio which can be fixed according to the hand-wheel setting. It can be seen that if the bottom pulley is driven by an engine and the top pulley is on the shaft of a threshing drum, we have a useful method of altering the threshing-drum speed. The concave hangs beneath the threshing drum and its position can be varied. Normally, it is used

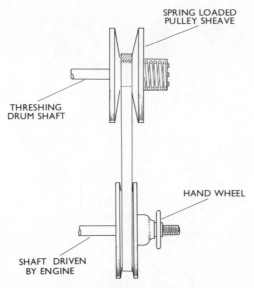

SPRING LOADED
PULLEY SHEAVE

THRESHING
DRUM SHAFT

HAND WHEEL

SHAFT DRIVEN
BY ENGINE

Fig. 151. A variable-speed pulley arrangement.

with a wider clearance at the front where the crop enters than at the rear where the straw leaves. Generally, it will be found that the larger grains and seeds are threshed with the wider concave settings whilst small seeds such as grass seeds require the concave to be set close to the drum. An instruction book for a given combine will state drum speeds and concave settings for any crop to be combined, but these can only be approximate because conditions of the crop can affect the ease with which they are threshed. For example, some varieties thresh out more easily than others; furthermore, much can depend on how dry the crop is.

To avoid damage to the grain and seed, it is always better to operate the combine with the drum running at the lowest possible speed and the concave set at the widest possible setting consistent with good threshing.

The Rear Beater

This is sometimes known as the stripper drum and it is situated immediately behind the threshing drum. Its purpose is to assist in the removal of the straw from the threshing drum and pass it on to the straw walkers.

The Straw Walkers

The straw walkers (Fig. 152) carry the straw out to the rear of the combine and deposit it on the field. It will be noted that they are slatted on the top of the box-like structure whilst this is also open at the lower end. The whole unit slopes upwards to the rear of the combine and, being mounted on a crankshaft, a rearward shaking motion is given to it. The type shown may be one of a batch of four but it is not unusual to have the straw walker as one complete unit occupying the full width of the combine cleaning area.

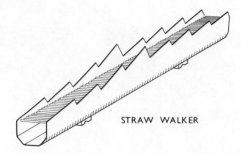

Fig. 152. A typical straw walker.

Any grain that comes out of the threshing drum with the straw should be shaken through the slats due to the action of the straw walkers and it should fall on to the straw walker bottom and slide back out of the open end. This grain then falls on to a grain pan.

The Grain Pan

The grain threshed out of the ears by the drum and concave, and the grain that comes from the straw walkers, falls on to the grain pan which is situated beneath the concave. This grain pan usually takes the form of a stepped plate occupying the full width of the cleaning unit and it also has a rearward shaking motion.

Whilst the grain falls through the concave and on to the pan there is also a large quantity of trash that comes through with it. Chaff, short ends of straw, weed seeds, etc., all come through mixed with the grain. This trash must be separated from the grain, so from the grain pan this mixture of trash and grain passes through a cleaning unit.

The Cleaning Unit

This consists of two sieves which also have a rearward shaking motion, and an air blast which can be directed up and through these sieves. The top sieve, often referred to as the chaffer sieve, extends from the grain pan and is adjustable so that the size of aperture can be increased or decreased. Grain and trash pass from the grain pan on to this sieve and the aim in setting this sieve is to set it at such a position as will allow all the grain to fall through it, whilst heavy trash is shaken rearward and out of the combine. If it is set too close, all the grain may not pass through and some may be shaken out over the rear of the combine, whilst on the other hand, if it is set too wide, excessive trash may go through it with the grain.

The bottom sieve, referred to as the grain sieve, is usually a sieve full of holes of a given size and is therefore not adjustable. However, this is not always the case and in some instances an adjustable sieve may be fitted. The purpose of the grain sieve is to carry out further separation of grain from trash should any trash have come through the chaffer sieve.

The grain sieve must be of such size as to allow the grain only to pass through and down to the grain auger. Anything larger will not pass through and will be shaken

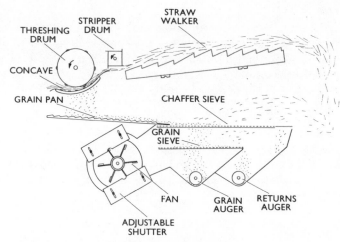

Fig. 153. The cleaning unit.

rearward to fall into the returns auger trough. Such material as broken heads will go to the returns auger.

Whilst the grain is moving over these sieves a constant blast of air is being blown up and through them. The purpose of this is to prevent the majority of the trash from settling on the sieves. By doing this, the sieves are kept clear and the grain can fall through readily.

The force of the air blown through can be varied and this is necessary because the weights of different grains and seeds vary. For example, the force of air required to keep the sieves clear when combining wheat would, when combining grass seed, also blow out the seed.

To alter the air blast, the usual method is to reduce the inlet through which the fan draws the air. This is done by adjustable blanking-off plates. In some instances it may be possible to alter the speed at which the fan rotates by fitting a different size of pulley wheel to the fan shaft, or by a variable-speed pulley.

Sieves and air blast must be set in conjunction with each other to suit a particular crop and wherever possible combining should be carried out with as much air blast as possible and with the sieves set as wide as possible.

This will ensure that little trash is allowed to settle on the sieves and anything lighter in weight than the grain is blown out. Also, the wide sieve setting will ensure that the grain can easily pass through.

The Augers

The auger beneath the grain sieve transfers the grain, which at this stage should be quite free of any trash, to an elevator which elevates it to either a grain tank or to a cleaning and bagging platform.

The returns auger transfers any broken unthreshed heads and the like to another elevator which elevates these to a position where they can either be returned to the threshing drum for rethreshing or put back over the cleaning unit. In some cases the returns may be taken to a separate unit entirely where rethreshing is carried out if necessary and the grain is then passed back over the cleaning unit.

The Grain Tank

Grain from the grain elevator is put into a tank built on to the combine and which is usually of sufficient size to hold between 1500 and 1750 kg (60-70 bushels). This usually amounts to about half an hour's combining. The tank can be unloaded into trailers either whilst the combine is stationary or working. This is done by a clutch-operated auger put in and out of drive by the combine operator.

Cleaner and Bagger

Where a grain tank is not used, a bagging platform is provided where a man is employed sacking the grain as it is being combined. Often, before the grain is sacked, it passes through a rotating screen which takes out any weed seeds that may have escaped the cleaning unit.

Additional Attachments

Various devices can be fitted to combines for use in different crops and crop conditions.

Grain lifters are used when the crop has been badly laid down by the weather and they help to lift it so that it can be cut and pulled in by the reel. These lifters fit on to the cutter-bar fingers and project 38-45 cm (15-18 in. approx.) in front of the cutter-bar, and it is usual to use one per foot length of cutter-bar.

Dividers of some form or other are normally fitted to the outside of the cutter-bar and their purpose is to push through the crop ahead of the cutter-bar, separating the crop which is being cut from that which is standing. These dividers can be fitted with extensions for use in laid crops.

Straw spreaders may be fitted to the rear of the combine to spread the straw across the field as it leaves the straw walkers. The spreader may take the form of a horizontal rotating disc positioned beneath the straw outlet.

Chopper spreaders may be fitted at the straw outlet, when straw leaving the combine is chopped into very short lengths and spread over the field.

Windrow pick-up attachments are used to pick up crops that have previously been cut by a mowing machine. This is often done to harvest such crops as grass seed, peas and trefoil seed. The crop is cut by a mower so that the seed can ripen in the swath and later the swath is picked up by this attachment which passes it into the combine for threshing.

Grain loss-monitors may be fitted to provide the operator with a guide to the combines performance. They consist primarily of acoustic chambers attached to the end of the straw walkers and sieves. Grain falling on these sensors has a percussion effect which creates an electrical impulse, this in turn is relayed to a control panel and displays a reading which compares with the original setting. Any increased reading shows an increased loss of grain and vice versa.

Routine Maintenance to the Combine Harvester

The combine harvester is another machine that requires careful and systematic lubrication attention. There are many greasing and oiling points, some of which will require attention more than once daily, others daily and some weekly. It is worth while for a combine driver to spend an hour each morning before combining, attending to lubrication, servicing and other maintenance jobs that need doing. If the combine is a self-propelled type, it will be powered by an engine and this will require the same sort of attention as does a tractor engine. Two points on the engine will require daily attention which may not always be given to a tractor engine. These are the air cleaner and the radiator. A combine works in very dusty conditions therefore the operator must ensure that neither of these become choked with dirt.

Each morning before work the combine operator should do the following:

(a) Remove any straw or trash that may have collected around drive pulleys or sprockets, etc.
(b) Remove any straw or trash that may have collected on parts within the combine, e.g. on sieves and straw walkers.
(c) Ensure that the grain pan and sieves are clean.
(d) Ensure that the bottoms of the straw walkers are clean. During damp conditions these can become choked with barley awns.
(e) Ensure that the concave is clear.
(f) Ensure that the threshing drum bars are clean. Dirt adhering to the bars can unbalance the drum and cause it to vibrate.
(g) Check the cutting mechanisms for adjustment at the knife head and for broken or damaged knife sections, etc.
(h) Check the tension of all drive chains and belts. This includes elevator chains.
(j) Service the engine and transmission system.

Throughout a day's work a combine driver is likely to have to make various adjustments to some of the mechanisms of the combine according to whether or not field conditions alter. The most likely adjustments that he will have to make will be to the drum, concave, sieves and air blast but it is the conditions alone that will determine what adjustment is made.

At the end of a day's work the combine should be covered with a tarpaulin.

Storing the Combine

The combine harvester is a very expensive machine, it can do a good job and do it at less cost and in much shorter time than can be done by other methods of harvesting seed crops. It is therefore worth while to care for it properly when it is working and also when it has finished the season's work. The life of a combine can be greatly reduced if it is not properly dealt with and left outside exposed to the weather when harvesting is finished. It is wise to adopt the following procedure at the end of harvest:

(a) Remove all straw and trash from outside and within the combine.
(b) Clean the grain pan, sieves, straw walkers, concave and drum of any dirt that may be adhering to them.
(c) Open the fan-blast shutters and elevator bottom doors. Start up the engine and put the threshing mechanisms into operation so that the dust and dirt is blown or transported out. Make sure that all grain or seed is removed from within the combine.
(d) Brush off any grain or seed from the cutter-bar bed, also from any other part of the combine where it may have accumulated.

(e) Remove drive belts if possible and store them in a dry cool place. If they cannot be removed, slacken off the tensioners.

(f) Remove drive chains and clean them thoroughly in paraffin, also clean the drive sprockets. Saturate the chains in lubricating oil and replace them.

(g) Remove the knife and clean it and coat it with anti-rust preparation.

(h) The engine should be cleaned down and given a good annual service. It would be wise to drain off the lubricating oil and refill the sump with a storage oil. When this is done the engine can be run for a few minutes to circulate this oil which will coat the inner mechanisms and prevent internal rusting. The cooling system must also be drained off.

(j) All parts of the machine that have worn bright should be coated with anti-rust preparation.

(k) All the lubrication points should be greased or oiled as required.

(l) The combine should be put in a good dry shed and the cutter-bar bed lowered on to blocks if necessary to keep it off damp earth. If it has to be stored in an open shed, it should be covered with a tarpaulin.

(m) If the combine cannot be jacked up to take the weight off the tyres, the tyres should be maintained at their normal pressure.

CHAPTER 26

Potato Crop Machinery

THE potato crop, as with the sugar-beet crop, can be an expensive labour-consuming crop if not mechanized, therefore, inevitably, it is now another crop that can be produced from start to finish almost entirely with machines. Automatic planters, various types of rowcrop cultivators, sprayers, potato spinners, diggers, tractor-hauled and self-propelled harvesters are the types of machines that are available for use in producing this crop.

It is not possible to deal here with the full range of machines and implements available, therefore only certain types will be discussed.

Within the range of different types of planters available there are those which have to be fed by hand, those that are not quite fully automatic and need an operator to feed potatoes to the planting mechanism when the mechanism has "missed", and there are the fully automatic types. These latter do not rely on any assistance to the planting mechanism.

The crop is grown in widely spaced rows ranging from about 610-900 mm (24-36 in.) and being so spaced allows ample room for the crop to develop, and for tractors with other implements and machines to travel between the rows to weed, spray and harvest the crop. The plants within the row may be spaced at anything from 250-450 mm (10-18 in.). These variations in row width and plant spacing within the row may be made use of by some farmers depending on their own ideas of how the crop should be grown, or some other specific requirement. For example, it is generally accepted that the closer the rows are together and the closer the potato seed within the rows, the smaller will be the potatoes when harvested. This may be an advantage if the crop is being grown for seed. If larger ware potatoes are required, then wider spacings will be used. The planter must be designed to allow for alterations in row and spacing.

The Hand-fed Planter

When potatoes are planted they may be in either a chitted (sprouted) or non-chitted condition. If they are chitted, they will require careful handling when being planted so as not to knock off the growing shoots; therefore, a planter that will not cause such damage is obviously better to use. However, if the seed are not chitted the need for an implement that will handle the potatoes with such care is reduced, provided of course that it does no other damage. Various types are available. A typical popular hand-fed planter which handles the potatoes with some care is shown in Fig. 154.

The illustration shows a two-row planter which is constructed for mounting to the hydraulic linkage of the tractor. This implement will plant either chitted or non-chitted seed. When chitted seed are to be planted, the sides of the hopper can if necessary be removed to allow the boxes of seed to be stacked on the bottom. By doing this, the

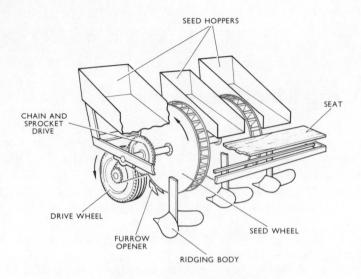

Fig. 154. A hand-fed potato planter.

seeds are not damaged by tipping them into the hopper and they are fed to the seeding mechanism from the boxes.

Operation of the Planter

This is simple yet positive. Refer to Fig. 154. The implement is designed to operate on the "flat", in other words, ridges are not previously prepared in which the implement will plant the potatoes.

As it is towed forward, two operators that are sitting on the seat, place a potato in each compartment of the seed wheels which are rotating forward in the same direction as the implement is travelling. The potatoes are carried around in the seed wheel and prevented from falling out of it by a fixed guard fitted to the frame of the implement and close to the outside of the wheel. This arrangement keeps the potatoes within the wheel compartment until they are near to the ground where the guard ends, and they fall a short distance into a furrow made by the furrow opener. The short fall distance is an added advantage when planting chitted seed because it reduces risk of damage to shoots.

The planting is completed when the ridging bodies on the rear of the implement cover the seed when forming the ridge.

Four main settings are provided on this type of planter to allow for different requirements. These are row width, seed spacing, planting depth and size of ridge.

1. Row Width

This can be altered by repositioning the seed wheels on the shaft. If the alteration is carried out it will also be necessary to reposition the furrow opener and the ridging bodies.

2. *Seed Spacing*

The seed wheel, which is the planting mechanism, is driven from the land wheel: therefore, the speed at which the seed wheel rotates will depend on the ratio between the drive sprocket on the land wheel and the drive sprocket on the seed-wheel shaft, plus the use of a constant forward speed in work.

By being able to fit different-sized sprockets to the land wheel it is possible, because the drive ratio has been altered, to alter the spacing at which the seed potatoes are placed within the row. The faster the seed wheel rotates, by using a small sprocket on the drive wheel, the closer the seed will be placed to each other, and vice versa. By using different-sized sprockets, seed spacings varying from about 230 mm (9 in.) to 450 mm (18 in.) can usually be provided.

3. *Planting Depth*

This can be altered by adjusting vertically the stem to which the furrow-opening device is attached. This device may take the form of a cultivator point fitted in front of miniature ridging bodies. The depth of the planting furrow should normally be between 50 mm and 100 mm (2 and 4 in.).

4. *Size of Ridge*

The size of ridge, and its shape, may be controlled by three factors. The depth at which the stem carrying the ridging bodies is set. The pitch of the ridging bodies. This can be controlled by adjustment to the top link of the tractor hydraulic linkage and will normally be set so that the implement runs level. The position of the mouldboards of the ridging bodies.

On most planters fitted with this type of ridging body it is possible to alter the ridge shape by making an adjustment to alter the distance between the mouldboards of the bodies. If the mouldboards of each complete ridging body are opened outwards, a high-crested ridge will be produced. The reverse affect, a flat-topped ridge, will be produced if the mouldboards are brought closer together.

An alternative to using ridging bodies to form the ridge over the planted seed is to use a concave disc assembly. This method is preferred by some farmers who claim that discs do not pack the soil tightly in the ridge as do some types of mouldboard ridgers.

Fertilizer Attachments

Most potato planters can be fitted with a fertilizer attachment so that the fertilizer can be sown as the potatoes are planted. The feed mechanism, which may be of the star-wheel type or an external force-feed mechanism, is driven from the implement's land wheel, as is the seed wheel.

The fertilizer is usually directed down coulter tubes to be placed in a band along the side of the potatoes and the application rate may be controlled by the use of different-sized drive sprockets on the drive shaft or driving wheel.

The Automatic Planter

Various types of automatic planters are available, some being fully automatic and some requiring occasional assistance from an operator riding on the implement. None of these types of planters are 100% accurate nor will they handle chitted seed with the care that a hand operator would: however, the speed of operation and the reduced labour costs in using such implements can often outweigh the disadvantages.   In probably all cases of automatic planters the use of well-graded seed can have considerable influence on the accuracy of the work done.

The essential features of a typical type of automatic planter are shown diagrammatically in Fig. 155. This type of implement can be provided to plant two, three or four rows at a time and may be arranged to be fitted to the tractor's three-point linkage system or trailed from the tractor draw-bar. A three- or four-row implement which is also fitted with a fertilizer distributor would require a fairly powerful hydraulic system to lift it when fully loaded.

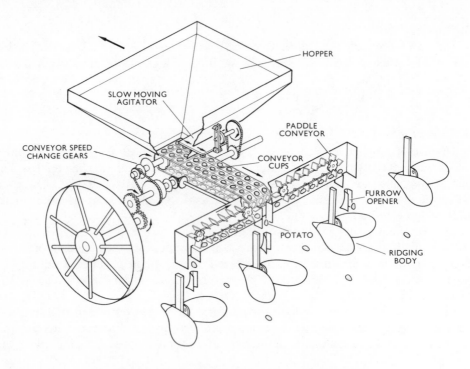

Fig. 155. Essential features of a typical automatic potato planter.

Operation of the Planter

This implement is designed to plant on the "flat". It is wheel driven and as it moves forward the various mechanisms operate to carry out the planting. Potatoes in the hopper feed into the cups of the moving conveyor, part of which forms the bottom of the hopper. This feed is regulated by a slow-moving agitator working in the hopper. Normally, when using well-graded seed, only one seed will be contained in each cup on the conveyor. However, should two small potatoes occupy a cup or should a cup be missed entirely, an operator riding on the implement rectifies the fault. A small tray containing potatoes is positioned close to him where he can take out or put in potatoes as required for the cup conveyor.

The cup conveyor carries the potatoes along towards the paddle conveyors and as can be seen from Fig. 155 potatoes in the centre line of cups fall straight to the ground, whilst those in the outside line of cups are conveyed by the paddles to the outside chutes and then fall to the ground.

The potatoes fall into a shallow furrow made by the furrow opener and planting is completed when they are covered by the ridging bodies.

As with other types of potato planters, the main settings for row width, seed spacing, planting depth and ridge size are provided for. Seed spacing, for example, is controlled by the speed at which the cup conveyor carries the potatoes to the paddle conveyors. Different speeds, usually three or four, can be provided by making use of changes in drive gear ratio. The forward speed of the implement when in use should be constant.

Routine Maintenance to Potato Planters

The more complicated the mechanisms of a planter, and this is likely to be more so with an automatic planter, the more care will be required to carry out satisfactory maintenance.

Before and during work the routine maintenance will consist mainly of lubrication. The row widths and seed spacing will be adjusted before starting work and the planting depth and coverage during the first few metres of work.

Lubrication attention will consist of greasing of grease points and perhaps the oiling of drive chains once or twice during the working day. There may be a need to limit the amount of oiling carried out on the chains if they are working close to the soil. If soil and grit get on to an oily chain the rate of wear to the chain and sprocket can be hastened because of the grinding paste qualities of the mixture.

When the planter has completed its season's work it should be given suitable attention before being put into store.

Firstly, it should be thoroughly cleaned down to remove all soil, etc.

It is fairly safe to assume that all parts of the implement that are polished due to normal working should be coated with some preparation that will prevent them from rusting. These parts will include mouldboards of ridging bodies, disc coverers and furrow openers.

Chains should be removed if possible, cleaned, oiled and replaced. Sprockets should be cleaned and oiled. All grease points should be given a shot of grease. If the implement is fitted with a fertilizer attachment, then this attachment must be given the attention that a normal fertilizer distributor should receive.

Finally, if possible, the implement should be kept under cover and, if fitted with inflatable wheels, jacked up on blocks so that the weight of the implement is off the wheels, otherwise the tyres must be kept to their correct pressures during the storage period.

The Potato Spinner

There are a number of systems or methods of harvesting potatoes and a number of different types of implements that can be used. The system adopted will be governed by certain factors, one of the main ones being the availability of labour. If sufficient hand labour is available and the acreage of potatoes grown not large, the potato spinner or elevator digger may be the implements used to get the crop out of the ground.

Three different types of potato spinner are available and in general these have been made for working in particular conditions. It is the soil type in which the potatoes are grown that usually determines the conditions. For example, an implement that will operate satisfactorily in light soil and expose the potatoes for hand picking could be fairly ineffective when used for the same job in heavy soil.

A popular type which works reasonably well in most conditions is shown in Fig. 156. This type of spinner may be in the form of a trailed wheel-driven implement, or a tractor-mounted p.t.o. driven implement. The latter is more popular because of its manouvreability and convenience in use and being power driven will work better in adverse conditions.

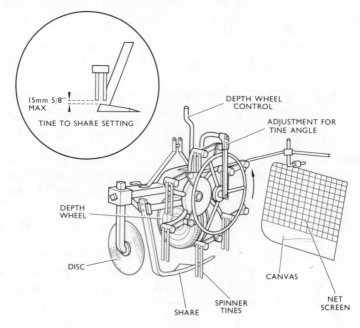

Fig. 156. A tractor-mounted potato spinner.

Operation of the Potato Spinner

Whatever type of spinner is used, the same general principles apply when it comes to setting and using it.

The implement should be attached to the tractor so that it lines up correctly with the row of potatoes to be lifted. This will normally be when the share point is positioned in the centre of the row. To achieve this it may be necessary to alter the wheel width setting of the tractor, or where possible reposition the implement on its main frame. One or two rows of potatoes will be straddled by the tractor depending on which setting is necessary and the make of spinner used.

From Fig. 156 it will be seen that the implement consists basically of a digging share and a set of rotating tines. It operates as follows. As the tractor moves forward, the digging share, which is full width of the row, travels along beneath the potato row and eases up and loosens soil and potatoes within the row. The rotating tines strike the row at right angles to the line of travel, and the soil and potatoes are spread over an area where the potatoes can then be picked up by hand.

It may be noted here that the action of the tines of this particular type of spinner is such that they push the row sideways rather than sweep radially through the row. This pushing action is achieved by having two wheels off centre to each other but connected by a linkage to which the tines are attached. The angle at which the tines work can sometimes be altered by adjusting the height of the outer wheel in relation to the inner wheel. The importance of this push action as opposed to a sweeping action is evident when damage to crop requires consideration. The former does less damage. On the other hand, in heavy field conditions the direct sweeping action of spinner tines is sometimes required to expose the crop—therefore, it is necessary to use such types.

As with any other farm implement the potato spinner needs to be correctly set to operate satisfactorily. There are a number of settings to be made.

1. *Depth of Work*

Assuming that the spinner is correctly positioned on the tractor in relation to the row, that is, with share point in the centre of the row, the depth should be set so that all the potatoes are lifted. Depth adjustment may be with depth wheel or hydraulic depth control depending on the make of implement. If the share is set too deep, excessive amounts of soil will be lifted and potatoes will again be covered by soil as they are spun out, whilst if it is set too shallow, many of the potatoes may be sliced and/or left in the ground.

Adjustment of the top link on the three-point linkage will have some affect on a mounted spinner because lengthening or shortening the link will alter the pitch of the share. Normally only slight downward pitch of the share will be required.

2. *Trash Disc*

The purpose of the vertical cutting disc is to cut away from the side of the row trash that may obstruct the spinner tines. It also has a stabilizing effect on the implement. The depth at which it is set is determined by its effectiveness in cutting through the trash.

3. *Speed of the Spinner Tines*

This should be as slow as possible consistent with good work. Excessive speed will not only damage potatoes but will result in them being spread over a greater area, thus making hand picking more difficult.

4. *Position of the Share in Relation to the Tines*

This is adjustable on some implements usually by adjusting the share vertically either on or by its stem. With the lower pair of tines positioned vertically and in a central position at the rear of the share, clearance between the ends of the tines and the top of the share should rarely exceed about 15 mm (5/8 in. approx.). A wide setting here could result in potatoes getting badly damaged, see Fig. 156.

5. *Position of the Screen*

The screen is used to prevent the crop from being spread in too wide a row. It is adjustable laterally and vertically and must be adjusted according to the conditions in the field.

Routine Maintenance to the Potato Spinner

Most spinners, whether the trailed wheel-driven type or the power-driven type, are fitted with a drive-gear housing in which the gears run in an oil bath. The oil used is normally a light gear oil which would normally require changing each season.

Apart from this the other lubrication attention will consist of daily or twice-daily greasing of grease points such as on the disc coulter bearing, land-wheel bearings, tine wheels and tine pivots. Share points will require replacement as wear takes place. If the spinner is power driven, its mechanisms will be protected by a slip clutch which must be correctly set.

When the spinner has completed its work for the season, it should be thoroughly cleaned down. The disc and share should be coated with anti-rust preparation, and all bearings should be given a shot of grease. If it is fitted with an inflatable wheel the pressure should be maintained and the implement stored under cover.

The Elevator Digger

The elevator digger is a potato-harvesting implement used in many areas especially where the soil conditions are not too heavy. It is a popular implement and works well in these conditions but in wet, sticky conditions the spinner is likely to be more effective.

One main advantage of the elevator digger is that it deposits the potatoes in a narrow row on the field and this eases considerably the work of the hand pickers. However, it is not reducing the overall labour requirements to any great extent because of the need to pick up the potatoes by hand whether they are in a wide or narrow row.

Compared with a potato spinner, there are many more components on the digger which come in contact with the soil and this inevitably means that more wear can take place. The extent of the wear will depend on the type of soil in which the implement is working. Sandy abrasive soils can cause high rates of wear to the various components.

Three types of the implement are produced, all of which are p.t.o. driven, and these consist of the trailed, semi-mounted or fully mounted types. Apart from these differences they are all operated basically in the same way. Of the three, perhaps the semi-mounted is the most popular. Any of them may be constructed to lift either one or two rows of potatoes at a time.

Operation of the Elevator Digger

Figure 157 shows a view of a semi-mounted elevator digger. The side is shown cut-away so that the working parts can be seen.

This implement is pulled from the lower linkage arms of the tractor hydraulic system which fit to the two attachment points, and is power driven from the tractor p.t.o. As with the potato spinner, tractor and implement must be so aligned that when the tractor wheel is positioned in between the two outside rows of potatoes, the digger share is cutting beneath the centre of the outside row. Tractor wheels will have to be

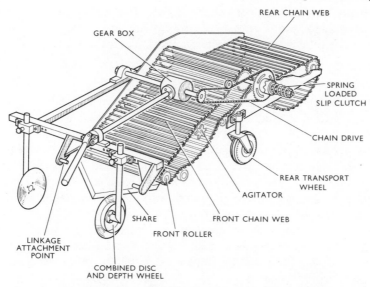

REAR CHAIN WEB

GEAR BOX

SPRING
LOADED
SLIP CLUTCH

CHAIN DRIVE

REAR TRANSPORT
WHEEL

AGITATOR

SHARE   FRONT CHAIN WEB

FRONT ROLLER

LINKAGE
ATTACHMENT
POINT

COMBINED DISC
AND DEPTH WHEEL

Fig. 157. Typical arrangement of components on a semi-mounted
elevator digger.

correctly set to achieve this. In some cases it may be possible to reposition the linkage attachment points to help in correctly lining up the implement.

From Fig. 157 it can be seen that there are many more components on the digger as compared to the spinner. The digging share performs the same function but in this case soil and potatoes are elevated over the top of the digger by the chain webs. The chain webs are more or less straight bars of steel linked together to form a chain and the bars are far enough apart to allow soil to pass through back on to the field. Therefore, as the digger is pulled forward potatoes and soil are separated by having soil go through the web and potatoes over the rear of the digger and placed into a narrow row on the field.

There are a number of settings and adjustments that can be made to an elevator digger so that it will operate satisfactorily in various conditions.

1. *Depth of Work*

The share should be positioned so that it cuts beneath the lowest potatoes in the row. This will be achieved by adjustment to the depth wheels at the front of the implement or by the hydraulic depth control where a fully mounted implement is used. Insufficient depth will result in potatoes being sliced through by the share whilst excessive depth will allow too much soil to be taken over. This latter situation can cause greater wear to components but on the other hand it should be remembered that soil is a softer cushion on which the potatoes can ride than is a number of steel bars, and less bruising of the potatoes may occur.

2. *Trash Disc*

Figure 157 shows a combined depth wheel and trash disc. On some diggers the two separate components may be fitted. The purpose of the discs is to cut away rubbish that may tend to obstruct the front of the digger. They can be adjusted vertically for depth and laterally to cut the sides of the row if necessary. This may be desired if it is thought necessary to reduce the amount of soil going over the web.

3. *The Chain Web*

The setting and arrangement of this is determined by the conditions in which the implement is working. A number of adjustments can be made to suit. Firstly, on most implements of this type it is possible to run the chain web as one continuous web or as two separate webs as shown in the illustration. Secondly, some of the rollers on which the chain web rides can be substituted with agitators and vice versa, if required.

Thirdly, the position of the bottom front rollers in relation to the share can sometimes be altered, e.g. raised or lowered. In addition to these three points the speed of the chain web can be varied by the p.t.o. speed.

With maximum numbers of agitators in use, with the chain web running in two sections, with the front rollers set to their highest position behind the share and with the web speed running fairly fast the implement would be set and working to provide maximum separation of potatoes and soil. This arrangement should only be necessary in the most difficult conditions.

The other extreme to this would be where conditions are very good and only the minimum of disturbance between potatoes and soil is required as the crop passes over the implement. In such a situation, no agitators would be used, the chain web would be run as one length, the bottom front rollers would be set in the lowest position and the chain web speed would be reduced. Forward speed may also be increased.

There are obviously many conditions in between these extremes where a combination of various settings on the implement are required to provide the desired result.

In setting the implement the following should be borne in mind:

(a) With the front rollers set higher than the share, soil separation starts immediately because the web is breaking the ridge of soil. Low setting of the rollers has the reverse affect.

(b) The more agitators used the greater the degree of separation of soil from potatoes, and the greater risk of damage to potatoes.

(c) The use of a double web increases separation, potatoes and soil drop from one web on to the other.

(d) Excessive working speed of the chain web will result in a greater degree of separation, and damage to potatoes.

(e) Potatoes riding on soil are less likely to be damaged.

(f) Increasing forward speed in relation to web speed will allow more soil to travel over the web. The reverse is also true.

(g) Aim at setting the implement so that the least possible damage is done to the potatoes and if possible they ride on soil up to the last foot or so of the web.

Some potato diggers are fitted with rubber-covered chain-web links as a means of preventing damage to the crop. Also an additional device can usually be fitted to the rear of the implement to deflect the potatoes to one side so that the implement can continue to work without having to wait for each row to be lifted before continuing.

Routine Maintenance to Elevator Diggers

The elevator digger works in conditions that are fairly severe and wear to the working parts of the implement can be high.

Lubrication attention should certainly be regular, it may be required two or three times during the working day but lubrication to the recommended points only should be carried out. Rollers and agitators are not likely to be provided with lubrication points because they are in continual contact with soil. These components are usually made of chilled cast iron to increase their wear resistance.

Wheel and disc bearings, power-drive couplings, slip clutch and main drive shaft bearings will all require attention plus the oil level in the gearbox.

The slip clutch will require setting correctly to prevent damage to the implement. When the season's work is completed, clean down the implement, renew worn parts, remove, clean and oil drive chains, coat the share with anti-rust preparation and grease all bearings. The gearbox can if necessary be replaced. Store the implement under cover if possible.

The Potato Harvester

If an implement such as the elevator digger can be designed and used to lift potatoes completely out of a ridge and separate them from the bulk of the soil, it seems folly to put these potatoes back on to the ground to be picked up by hand. The logical solution is to develop and extend the implement in such a way as to transfer the crop into some container or other, be it sack, crate, or trailer, and avoid having to pick it up again off the field. This is in fact what is happening when a potato harvester is used because the harvester is in most cases an extension of the elevator digger. There are, of course, types of potato harvesters which operate on entirely different principles to the elevator digger. All types aim at lifting the potatoes, complete with soil, out of the ridge and by passing them through different sections of the implement separate them from soil, stones, tops and any other rubbish. The potatoes may then be directed to either sack, crate or trailer depending on the system adopted.

Figure 158 shows the layout of the mechanisms of a typical potato harvester and the path of the crop through it. Only the main working parts are shown.

The implement is trailed and power driven by the tractor hauling it. As it moves forward along the row, the steel discs cut through any trash that may tend to block up the mouth of the elevator. These discs are adjustable vertically and laterally, and may be set to cut soil away from the ridge sides to reduce the amount taken up the elevator.

Two main problems which exist when harvesting potatoes with an implement of this type are the separation of potatoes from stones and clods, and damage to the potatoes. Therefore where possible, throughout the implement provision is made to try and lessen these problems. The disc setting to reduce soil intake may be necessary on some types of soil.

The digging share is usually in a fixed position in front of the elevator and attached to the elevator frame. It is adjustable vertically and adjustment of it also raises or lowers the elevator at the intake end. As with the elevator digger, the digging share should be set so that it cuts beneath the lowest potatoes. Different types of shares are usually available for fitting in place of a standard type where conditions require it. For example, some shares are designed to reduce soil intake on to the elevator.

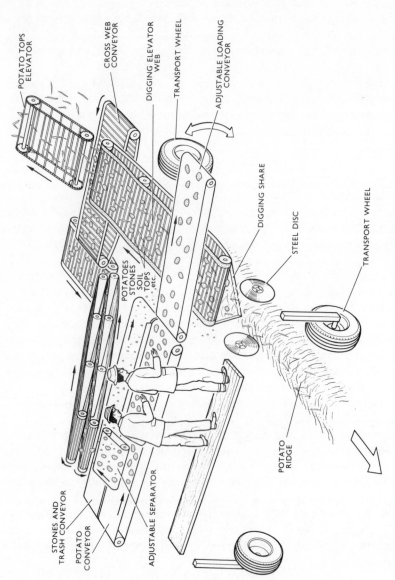

POTATO TOPS
ELEVATOR

CROSS WEB
CONVEYOR

DIGGING ELEVATOR
WEB

TRANSPORT WHEEL

ADJUSTABLE LOADING
CONVEYOR

POTATOES
STONES
SOIL
TOPS
etc.

DIGGING SHARE

STEEL DISC

TRANSPORT WHEEL

STONES AND
TRASH CONVEYOR

POTATO
CONVEYOR

ADJUSTABLE SEPARATOR

POTATO
RIDGE

Fig. 158. Layout of the main components of a potato harvester.

As the share is moving through the ridge beneath the potatoes its action is to start breaking up the ridge completely so that soil and crop, etc., then pass on to the elevator web. It is when the mixture of soil, stones, tops and potatoes, etc., gets on to this elevator that the separation of potatoes from anything that is not a potato begins. The elevator web is a series of metal bars spaced apart and rivetted to belting to form an endless conveying system moving rearwards. The spacing of the bars allows soil, small stones and only the smallest potatoes to fall back on to the field.

At the top of the digging elevator web the tops are elevated off by an elevator fitted with bars spaced at wide intervals so that whilst tops are carried on the elevator, potatoes will fall through the bars on to the cross-web conveyor. This conveyor is of a similar type to the digging elevator but may be fitted with rubber-covered bars to prevent bruising of the potatoes as they fall on to it. In fact all other bar elevators may be so fitted. Some clods falling from one conveyor to another may tend to break up and pass through the bars.

From the first cross-web conveyor it can be seen that the potatoes, etc., are then transferred to two other web conveyors before coming in contact with an adjustable separator. The double-web conveyor which conveys the potatoes, etc., between two webs can eliminate many clods in some soil conditions merely by the pressure applied to them by the web bars. This pressure is not great enough to damage potatoes nor to crush the harder clods.

The adjustable separator is in effect an endless belt rotating in the opposite direction to that which the potatoes, etc., fall on to it. It can be lifted or lowered at the end nearest to the potato conveyor. Potatoes roll down it on to the potato conveyor whilst flat-sided or rough-sided objects tend to settle on it for long enough to be carried over and dropped on to the stone and trash conveyor. Altering the angle of the separator affects the degree of separation of clods, etc., from potatoes and therefore the amount of each going on to the two conveyors. Its angle must be set to suit the conditions.

A number of operators, usually three or four, work at these conveyors, there may be two workers on each side and they are occupied in removing stones, etc., from the potato conveyor and potatoes from the trash conveyor, and putting them where they should be.

Finally, the potatoes are transferred to a loading conveyor which conveys in this case to a trailer travelling alongside the harvester and stones and trash separated are conveyed back on to the field. The loading conveyor is adjustable for height so that the potatoes may be delivered into the trailer always from a low position so as not to damage them as they drop, which will happen if they are dropped from a high conveyor.

Routine Maintenance to Potato Harvesters

This can vary somewhat depending on the type of potato harvester but certainly the maintenance should be thorough because, as with the elevator digger, the potato harvester is working in fairly severe conditions.

If a gearbox is fitted it must be checked for oil before work commences, the oil changed if necessary and maintained at its correct level. There can be many greasing points on an implement of this type, these will require greasing at least twice throughout the working day but only the operator's instruction book can give the correct information. Belt and chain tensions must be attended to. Slip clutches must be properly set. Conveyors and elevators properly adjusted. Tyre pressures correct.

Thorough maintenance and servicing will result in fewer stoppages due to breakdown when the implement is at work.

At the end of the season the potato harvester requires similar attention to many other farm implements. Clean the implement down thoroughly. Replace worn parts. Slacken belt tensions or remove them to indoor storage. Elevator webs and conveyors may also be removed and stored indoors. Grease all bearings. Apply anti-rust preparation to the share and disc coulters.

Maintain tyre pressures. Store the implement under cover.

CHAPTER 27

Sugar-beet Crop Machinery

OVER the past years much work and development has gone into producing implements and machines that will reduce the amount of labour required to grow the sugar-beet crop. It is now almost possible to completely mechanize the production of this crop from start to finish and thus considerably reduce the number of man hours entailed as well as making the work that is done much less laborious than it used to be.

The crop is grown in widely spaced rows 45-54 cm (18-21 in. approx.) apart. Within the rows, the plants are normally singled or chopped out to leave a plant about 23 cm (9 in. approx.) from its neighbour. This arrangement allows ample space in which each plant can fully develop and provides a suitable width between plant rows down which tractors and other equipment can travel to weed, spray and harvest the crop.

The greatest hand labour requirement that has not yet been completely eliminated when growing this crop is that of singling and gapping the beet plants to leave plants standing in the row a distance of about 23 cm (9 in. approx.) from each other. Machines are available which will reduce the number of plants in the row by knocking them out of the row, but will not leave them evenly spaced. Nevertheless, the use of this type of machine, known as a Down-the-row Thinner, will help reduce the amount of time that hand labour has to be used to complete the job.

A further aid to reducing this hand labour is to sow the crop with a seed drill that will place the seed at a specific distance apart, sometimes called the precision seed drill, thus greatly reducing the number of plants that will actually grow in the row. In establishing a crop with this type of drill, one of two methods may be adopted. The traditional method would be to drill the seed on 45-55 cm (18-21 in.) rows and at a spacing of 50-75 mm (2-3 in.) approximately. After allowing the crop to germinate it would then be "gapped" and "singled" to leave one plant every 23 cm (9 in.) approximately. The reason for this being that sugar-beet seed is naturally a multi-germ seed which will produce more than one plant therefore singling is necessary. Also the germination rate is poor, around 70%, and this means over-sowing initially and after emergence "gapping" the crop to the desired population of 74,000 plant/ha (33,000 plants/acre).

The second method is to drill to a stand. At the present time in Gt. Britain this method is practised in some areas by as many as 75% of sugar beet growers, whilst in some other areas about 35% practice it. The efforts of plant breeders has now led to the production of monogerm sugar-beet seeds, eliminating the need for singling, and this in turn has encouraged the farmer to sow the seed on 50-cm (20-in.) rows at a spacing of 15-16 cm (6¾ in.). Allowing for the germination rate this produces the correct plant population and had eliminated the two laborious and costly operations.

If a precision seed drill is not used then a type known as a continuous-flow drill will be used. These drills are used much less so than they used to be, particularly for sugar-beet seed, but they can and are used for sowing a variety of other types of seed. In operation they sow a continuous line of seed, thickly or thinly depending on the setting made on the drill and of course numerous plants grow close to each other in the row. A typical drill of this type is in the form of an independent single seeder unit.

The Continuous-flow Seeder Unit

When sugar-beet seed is sown it is usually done so that not less than four rows of seed are sown in one bout across the field, therefore, when independent seeder units are used, a number of them, four or more, will be attached to a tool-bar or frame work to be towed by the tractor. Usually the complete set on the tool-bar is carried on the hydraulic lift linkage of the tractor. This makes it easy to transport and operate in the field.

A diagram showing the main working parts of a typical continuous-flow seeder unit is shown in Fig. 159. When this type of unit is attached to a tool-bar it is done so in

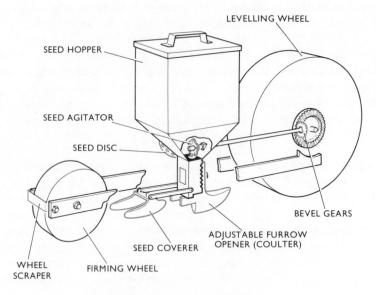

Fig. 159. Continuous-flow seeder unit.

such a way that the unit is free to rise and fall with undulations on the field. Some form of spring pressure is also applied to it so that the unit is also always being firmly pressed down on to the field surface.

The Hopper

The seed hopper is of small capacity usually holding between 2 and 4 kg (3½-8½ lb approx.) of sugar-beet seed. The larger capacity hopper would therefore carry enough seed to sow about ·2 hectare (½ acre), therefore, with four units on a tool-bar, one filling of all the hoppers would carry enough seed to sow ·8 hectares (2 acres).

The Sowing Mechanism

The sowing mechanism is simple yet positive, and consists of a revolving metal disc or agitator which is waved so that when it rotates it pushes seed through an aperture of a given size situated in the hopper bottom. This agitator is driven from the front land wheel of the unit, see Fig. 160b. Other types of sowing mechanisms may be used in this type of machine, one such being in the form of a circular brush; however, this type is prone to wear and, when the bristles wear off the brush, the seeding rate is affected.

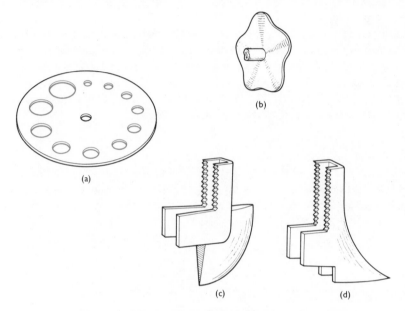

Fig. 160. (a) Seed disc. (b) Wavy disc. (c) Cast iron coulter. (d) Steel coulter.

The Seed Disc

Combined with the agitator, the seed disc controls the sowing rate of the unit. Figure 160a shows a seed disc and it can be seen that it is simply a flat circular metal disc which has a number of different sized holes around its circumference. Any one of these different sized holes can be held in a fixed position over a main aperture in the hopper bottom. The agitator is revolving above this main aperture. Depending on the size of hole positioned over the aperture, so the seeding rate is governed; for example, a large hole will allow more seed through than a small, and so on, and it is only necessary to position the correct sized hole over the aperture to get a desired sowing rate. A seeding unit of this type will be provided with a number of seed discs, each having a different range of hole sizes so that a very wide range of seed types and rates of sowing can be catered for. Seed types varying from cabbage to beans can be sown.

The Furrow Opener (Coulter)

When the seed passes through the hole in the seed disc it falls straight down to earth through a short tube and into a furrow made by a furrow opener. The depth at which the seed is sown is determined by the setting of this furrow opener. From Figure 159 it will be seen that it is adjustable and a fairly accurate method of setting it correctly on a set of units carried on a tool-bar is as follows: with tool-bar and seed units mounted on the tractor hydraulic linkage, stand the tractor on a level concrete surface. If, for example, the sowing depth required is 25 mm (1 in. approx.) then place two wooden boards of this thickness across the width of the tool-bar and seed units, one board being set beneath the front wheels of the seed units and one board beneath the back wheels. Lower the implement on to the boards and then adjust each furrow opener to just touch the ground.

Different types of furrow openers (coulters) are available for different field conditions. The type shown in Fig. 159 is a cast iron coulter used for lighter soils where a good seed bed can be obtained. Figure 160 shows two other types. Figure 160c is another cast-iron coulter known as a 'Suffolk' coulter and is used where there may be considerable trash in the seed bed. Figure 160d is a steel coulter used on heavier land to get better penetration or where greater depth of sowing is required.

The Seed Coverer

The seed coverer trails on the field surface behind the coulter and its purpose is to cover the seed furrow with soil, thus burying the seed. Various types of these are in use, some being adjustable to vary the amount of soil pulled on to the furrow. The one shown in Fig. 159 is not adjustable.

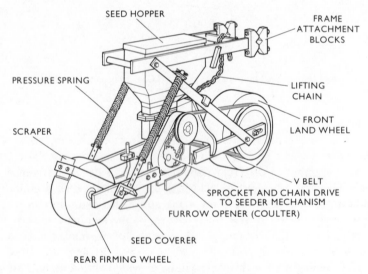

Fig. 161. A precision seeder unit.

The Seeder Unit Wheels

Whilst it may be necessary to have wheels on these units in order that they may travel across the field, the wheels also serve other purposes. The front wheel is driving the seeder mechanism, but not only this, it also levels and firms down the seed-bed directly ahead of the furrow opener and because of this the furrow opener may make a more positive cut in the soil.

The rear wheel travels on top of the covered furrow into which the seed is sown and gently consolidates the soil around the seed.

Precision Seed Drills

Sowing the crop at accurate spacing calls for the use of a precision drill which is now used extensively not only for sugar-beet production but also for such crops as cauliflower, turnips, cabbage, etc.

These drills have the advantage in that they will place single seeds accurately at set distances apart. Also, if required, some types of drills will sow small groups of seed at set intervals, for example, three seeds every 15 cm (6 in. approx.).

Individual drill units (Fig. 161) are attached to a tool-bar frame. The number of units being determined by the harvesting system in so much as the total number must be a multiple of the harvesters lifting capacity, (i.e. 3 row harvesting system: 3-, 6-, 9- or 12-row drill unit). With a single-row harvester any drill unit size can be used.

A seed spacing drill is similar in shape and size to a continuous flow drill. It has a small hopper, adjustable furrow opener, seed coverer and two land wheels. A land wheel drives the seeding mechanism which meters the seed into a furrow made by the furrow opener, this is then covered by the seed coverers and the soil is compacted around the seed by means of the rear firming wheel.

The Seeding Mechanism

As shown in Fig. 162 the seeding mechanism consists of two wheels and a seed ejector. The wheel that does the actual sowing of the seed has in this case got a series of holes drilled at specific distances apart around the circumference of the wheel. These holes are of such size and depth as to allow one seed only of a particular type to fall into the hole. For example, for drilling sugar-beet, the holes, which may be

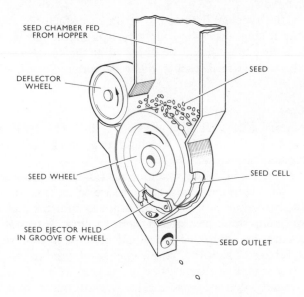

Fig. 162. A precision seeder mechanism.

called seed cells, may be of such size as to accommodate a single seed between 2·77 mm and 4·366 mm (7/64 in. and 11/64 in.) diameter. For drilling cabbage the holes may be large enough to accommodate single seeds between 1·956 mm and 2·108 mm (0·077 and 0·083 in.) diameter. This means, of course, that each different type of seed being sown requires a seed wheel to match the seed.

The seed wheel is rotating in the hopper bottom and is in direct contact with the seed and, as a seed occupies each of the cells, it is carried around to the ejector plate. The seed cannot get out of the cell until it reaches the ejector because it is held in position by the frame of the mechanism which hugs the wheel circumference.

In addition to the cells, a thin groove is cut around the wheel and it is in this groove that the ejector is held in a fixed position so that all seeds are carried around and come in contact with it. The ejector is so shaped that it causes each seed to come out of the cell and at this point the seed can fall freely through a seed outlet and to the ground. A suitable type of ejector may be required to eject a particular type of seed.

The purpose of the deflector wheel is to push back any seeds other than those in the cells of the seed wheel and this is achieved by having it rotate in the same direction as the seed wheel. The deflector wheel may be rubber covered to prevent seed damage or made of light alloy with a lightly serrated face around its circumference.

Seed Spacing

The distance at which the seeds are placed apart in the furrow depends mainly on two factors:

(a) the distance at which the cells are placed apart around the seed wheel and
(b) the speed at which the drill unit is used in the field.

By using suitable seed wheels it is possible to sow single seeds at distances apart varying from about 39 mm to 30 cm (1½-12 in. approx.) and of course many different types of seed can be sown.

Seeder units of this type are designed to operate satisfactorily at a speed of about 4 km.p.h. (2·5 m.p.h.) but, depending on the type of unit, it may be more or less. The important thing is that they should be operated at the speed recommended by the manufacturer otherwise the seed spacing will be inaccurate.

The Complete Unit

A typical precision seeder unit is shown in Fig. 161. The drive in this instance is from the front land wheel via a V-belt which drives a sprocket and chain and which in turn is driving the seed mechanism.

The unit is arranged for attaching to a tool-bar frame. Adjustable pressure springs are fitted to maintain a constant downward pressure on the rear firming wheel. These should be adjusted to apply just enough pressure to firm the seed in satisfactorily.

The front wheel is adjustable on a slotted frame to allow for tensioning of the V-belt. The seed coverers are also adjustable and should be set so as to move just enough soil to cover the seed furrow. The wheel scraper may be adjusted to suit conditions but its purpose is to prevent the rear wheel becoming stuck up with soil. The lifting chain is attached to the frame so that the unit hangs downwards at the front thus allowing the front wheels to contact the ground first when they are lowered into work.

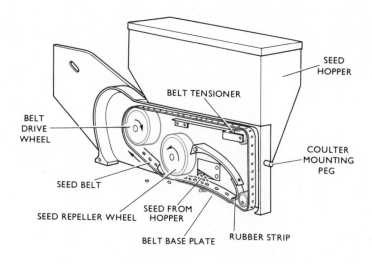

Fig. 163. A belt-feed precision seeder mechanism.

The cell wheel type of precision seeder unit is fairly widely used but another type that is widely used makes use of an entirely different method of sowing the single seeds. An illustration of the seed mechanism of this type is shown in Fig. 163 where it can be seen that it consists of a rubber belt with a series of holes in it to accommodate the seed which is carried along beneath a repeller wheel and the seed falls through the hole to the ground.

The seed feeds through an aperture from the main hopper to the chamber and the amount of seed fed into the chamber is regulated to prevent over-feeding which may cause choking and wear to the mechanisms. Regulation of the feed is achieved by the use of a special choke plate and, as this type of drill will sow many different types of seed of various shapes and sizes, it is necessary to fit a suitable choke plate which will provide a satisfactory flow of seed into the chamber.

The seed belt and the metal base on which it runs are also designed to cope with seeds of a specific type and size, therefore it will be necessary to fit different types when different types of seed are to be sown, for example, when changing from sugar beet to cabbage seed.

The rest of the seed drill is basically similar to the other types already dealt with.

The Multi-unit Assembly

It is usual for four or more to be used on a tool-bar for drilling and for accurate work the units must be accurately positioned on the tool-bar. The tractor wheels must also be correctly set so that they are not running on land immediately in front of a seeder unit. The wheels should be set at a distance apart, measured from centre to centre of the wheels, equal to a multiple of the row width. For example, if the beet is to be sown in 50-cm (20-in.) rows then the tractor wheels will have to be set at 150-cm (60-in.) centres. Figure 164 shows the arrangement and setting for a five-unit outfit.

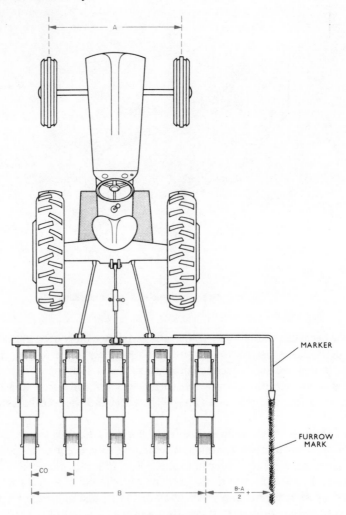

Fig. 164. A five-seeder unit arrangement with marker setting.

Seed-sowing Indicator

When an operator is using precision seeders in the field he cannot see the seed being sown. The units are too small, too close to the ground and too far away from him for him to see anything more than the land wheels revolving. It is conceivable, and in fact does happen, that something goes wrong within a unit and it ceases to sow its seed. If this happens, the operator must know immediately otherwise he may well drill a few acres of land before discovering it. To help avoid such a situation, electrical indicator controls can be fitted to the tractor and seeder units and these indicate by flashing lights whether or not the seeder units are operating satisfactorily. For example, each unit is electrically wired to a main indicator panel which is situated directly in view of the driver. Each seeder unit may have a contact-breaker device within it which is driven by the land wheel of the unit, and as this makes and breaks contact in the electrical circuit to the unit, the light on the indicator panel will flash intermittently.

Should the flashing cease, for example, either there is no light at all or a continuous light, indicating that the contact breaker has ceased to operate and the contacts are either continuously open or closed, then the tractor driver must investigate the cause.

Markers

To get accurate spacing between the rows sown in one bout across the field and the next bout, it is necessary to have an accurately set marker. This is more important on a set of seeding units of this type than on a grain drill because any variation between row widths can result in plants being chopped out of rows which are the wrong distance apart, when inter-row cultivations are carried out.

Setting of a marker has been dealt with in Chapter 16 and the same method can be applied to setting the marker when using a set of seeder units. Referring to Fig. 164, for example, if:

Row spacing (coulter width)	=	50 cm (20 in. approx.)
Overall width "B" (coulter to coulter)	=	200 cm (80 in. approx.)
Tractor-wheel setting "A"	=	150 cm (60 in. approx.)

$$\text{Then marker setting from outside coulter} = \frac{B-A}{2} + 1 \text{ coulter width}$$

$$= \frac{200 - 150}{2} + 50$$

$$= \frac{50}{2} + 50$$

$$= 25 + 50$$

$$= 75 \text{ cm (30 in. approx.)}$$

Master Land-wheel Drive

The master land-wheel drive implement which consists of a pair of large-diameter heavily treaded wheels driving all the seeder units on the tool-bar has the advantage that it provides a positive drive. Wheelslip which may affect seed spacing is eliminated. Usually where this type of drive is used it is also possible to alter the drive ratio to the seeder units, either by use of a gearbox or various belt pulleys in the drive. By being able to do this a greater range of seed spacings may be obtained for a given seed-spacing wheel.

Routine Maintenance to Precision Seeder Units

There are not a lot of lubrication points on a precision seeder unit, there may in fact be only two, one on each land wheel, but of course these must have their daily greasing. Belt drives must be kept properly tensioned to prevent wheelslip. Proper care when using the seeder units will be worthwhile in that better results will be obtained and the main points to note are as follows:

(a) The implement is designed to do its job accurately at a recommended forward speed. This speed which may be 3·2-4·8 km.p.h. (2-3 m.p.h.) should be maintained.
(b) The mechanisms in the seeder unit are designed to carry the seed on one direction, do not reverse the drive otherwise damage may be done.

(c) The mechanisms in the seeder unit are designed to sow graded seed and seed only. See that nothing other than the graded seed gets into the hopper.

(d) Do not clog up the seed outlet with soil. This can be avoided if the seeders are not reversed whilst in work and if the seeders are lowered gently into work as the tractor is moving forward.

(e) Dust, seed dressing, damaged seed, large seeds, etc., will tend to clog up the seed chamber and seeder mechanisms therefore these should be cleaned out at least daily whilst in use.

(f) The repellor-wheel rubber tyre where fitted may require replacing if it has worn grooves otherwise seed spacing may be affected.

At the end of the season, thoroughly clean the implement, clean out the seed chamber and seeder mechanisms, making sure that all seeds and corrosive seed dressings are removed and store the implement in dry conditions.

The Sugar-beet Harvester

The development of the sugar-beet harvester has been so successful that few farmers who grow sugar beet are without a harvester of some sort. Even farmers who grow quite small acreages of this crop seem to justify having a harvester of their own.

Without a harvester the lifting of the crop can be a laborious time-consuming job, although simple toppers and lifters are available which may ease the work considerably. It is from these simple toppers and lifters that the present-day harvesters have developed. Most of the beet harvesters available to the farmer are p.t.o. driven types which all operate in basically the same way, but self-propelled types are also now available. These latter machines have built-in self-unloading tanks which can eliminate the use of an additional tractor and trailer for carting the sugar beet. Even so, these self-propelled machines operate in much the same way as the other types and make use of the same types of lifting and topping devices.

Briefly, the complete operation done by a beet harvester is as follows:

A pair of lifting shares or wheels lift a previously topped row of beet on to a chain elevator which carries the beet to a cross elevator which in turn delivers the beet into a trailer running alongside the harvester. Where a self-propelled tanker machine is used, this elevating arrangement is modified.

Whilst the lifting of one row of beet is taking place, another row is having the tops cut off and the tops are swept to one side so that the row is clear for the lifting share on the next time round.

The Topping Device

To get maximum crop weight when the sugar beet is harvested it is necessary to top the beet correctly, that is, remove just sufficient of the crown of the root to get rid of all the green leaf growth. If the sugar beets are consistently over-topped by cutting off too much of the root, a high percentage of the crop is wasted and on a large acreage of beet this wastage can amount to many kilogrammes. On the other hand, if the sugar beets are consistently under-topped, and a lot of green leaf is left on the crown of the beet, the grower will be penalized by the beet factory by having the total amount of excess leaf and crown per kilogramme estimated, and deducted from the total weight sent to the factory. It is, therefore, an advantage to both the grower and the factory if the sugar beet is correctly topped.

The modern beet harvester is fitted with a topping mechanism which, if correctly set, will satisfactorily top the beet and it is due largely to the success of this device that beet harvesting is no longer a serious headache to the farmer.

The topping mechanism is usually driven by one of the land wheels of the harvester and it consists of a spring-loaded feeler wheel and a horizontal knife. Figure 165 shows

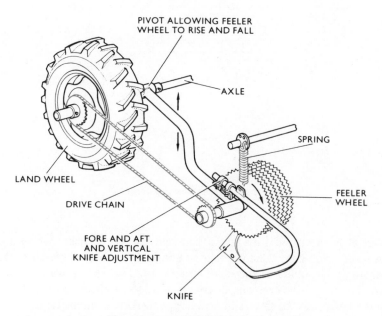

Fig. 165. A typical topping unit arrangement.

a typical arrangement of the drive and topping mechanism. It is important that this mechanism is correctly set and there are a number of adjustments that can be made to it so that it can be set correctly to suit the crop. Firstly, when the harvester wheels are set so that they run in between rows of beet, i.e. if the beet is growing in 50-cm (20-in.) rows, the harvester wheels should be set at 150-cm (60 in. approx.) centres, as should be the tractor wheels, the feeler wheel should be positioned centrally over the crown of the beet. There will be provision on the harvester to allow for this lateral adjustment of the feeler wheel.

Secondly, the whole feeler wheel unit, plus knife, is carried on a tension spring which in effect allows the unit to float. Adjustment of the tension of this spring has the effect of either making the feeler wheel ride heavily or lightly on the crop. The adjustment must be made to suit conditions, bearing in mind that if the beet tops are bulky the tension on the spring should be reduced so that more weight of the feeler wheel is on the top of the beet. This is necessary because the wheel helps to hold the beet in position whilst the knife cuts through the crown. The tension can be increased if the tops are light but at all times the feeler wheel must be allowed to float so that it can rise and fall to suit the various heights of the beet.

Thirdly, the position of the knife in relation to the feeler wheel determines how the beet will be topped and how much crown will be removed. The knife can be raised or lowered but its final work position will depend on the nature of the beet. As a general guide a clearance of between 7 and 10 mm (1/4-3/8 in. approx.) will be a reasonable setting to start with, see Fig. 166.

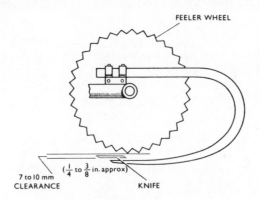

Fig. 166. Knife to feeler wheel clearance.

The knife can also be moved fore and aft in relation to the feeler wheel and its position in this respect will determine whether or not the crown of the beet is sliced off straight or at an angle. It should be sliced off straight and Fig. 167 shows the effects of various settings. The size of the beet will, of course, have some effect on this. Therefore the position of the knife must be set to suit an average size of beet in the field.

The knife must also be kept sharp to do good work and this can be done either by a file or by an emery wheel. It is sensible to carry a spare knife to avoid holding up the work for too long a period when a knife becomes blunt.

The Top Flail

When the tops of the sugar beet are cut off by the knife they lie on the row of beet where they are cut and it is necessary somehow to move them away, so that the lifting mechanism is not impeded by the tops and tops are not elevated with the beet.

A flail device is invariably used for this purpose and it may be positioned either at the rear of the topping unit or immediately ahead of the lifting mechanism and main elevator. In construction it takes the form of a chain-driven or power-driven steel disc fitted with rubber flails. Figure 168 shows an illustration of a flail device. Correct setting of it is important and usually both vertical and horizontal adjustment is provided for it on a beet harvester. The setting must be related to the crop and a normal setting is such that the rubber flails just flick the soil as they sweep across the row and if they are also set about 25-50 mm (1-2in.) to one side (see Fig. 168) they will be less likely to strike the beet. This should be avoided if possible, especially if the beets are growing well out of the ground, otherwise the flails will tend to knock the beet out of the ground. The vertical adjustment for the flail must be correctly set to avoid this.

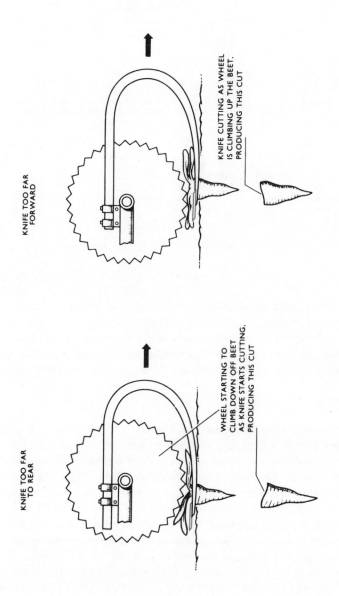

KNIFE TOO FAR
TO REAR

KNIFE TOO FAR
FORWARD

WHEEL STARTING TO
CLIMB DOWN OFF BEET
AS KNIFE STARTS CUTTING.
PRODUCING THIS CUT

KNIFE CUTTING AS WHEEL
IS CLIMBING UP THE BEET.
PRODUCING THIS CUT

Fig. 167. Faulty knife settings.

On some harvesters it may be possible to increase the speed of the flail by substituting different-sized drive sprockets and this may be necessary where there is a lot of top growth on the beet and excessive weed in the rows.

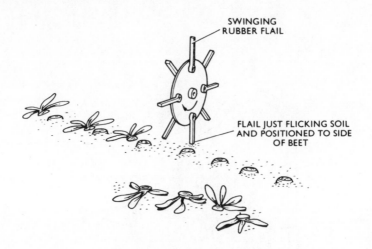

SWINGING
RUBBER FLAIL

FLAIL JUST FLICKING SOIL
AND POSITIONED TO SIDE
OF BEET

Fig. 168. Position of tops flail.

The Disc Coulter

Most beet harvesters are fitted with a disc coulter which may be flat or concave, and its purpose is to run ahead of the topping unit and clear leaves and other growth from the path of the heel of the knife thus preventing any blockage around the knife. The coulter is adjustable for depth and angle and should be set and angled just sufficiently to clear any trash and produce a shallow furrow in which the knife heel can trail.

The Lifting Shares

Perhaps the most common form of lifting device used on beet harvesters is a pair of triangular-shaped steel shares which are situated immediately ahead of the main elevator. They are so positioned that as the beet comes out of the soil it passes straight on to the elevator for ultimate discharge into a trailer or tank. Figure 169 shows an illustration of typical shares and there are a number of adjustments provided on them so that they may be satisfactorily set to suit conditions in a given crop. For what may be termed average conditions, that is, neither too wet nor too dry, the shares should be set so that the bottom edge is level with the ground. It should be noted that these shares are not digging the beet out but pressing the soil downwards at the sides of the beet and this has an effect of oozing the beet out of the ground. As the harvester moves forward and with a share on either side of the beet, the beet rises up the shares and on to the elevator.

Too much downward pitch on the shares will cause varying amounts of soil to come up with the beet, particularly when the land is wet and heavy. Therefore in such conditions it is advisable to pitch the points of the shares upwards. However, excessive upward pitch of the share points can result in fangs of the beet snapping off.

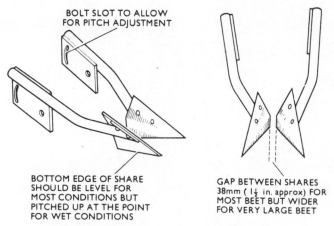

BOLT SLOT TO ALLOW
FOR PITCH ADJUSTMENT

BOTTOM EDGE OF SHARE
SHOULD BE LEVEL FOR
MOST CONDITIONS BUT
PITCHED UP AT THE POINT
FOR WET CONDITIONS

GAP BETWEEN SHARES
38mm (1½ in. approx) FOR
MOST BEET BUT WIDER
FOR VERY LARGE BEET

Fig. 169. Typical arrangement of beet-lifting shares.

The depth at which the shares are allowed to penetrate into the soil is governed by the setting of a pair of depth wheels at the front of the machine or in some cases by an hydraulic depth controller. Generally a penetration of about 5 cm (2 in.) approximately of the shares into the soil will be satisfactory but this is a field setting that must be adjusted to suit conditions.

The gap setting between the heel of the shares is largely determined by the average size of the beet. More often than not a gap of about 38 mm (1½ in. approx.) is satisfactory but if the beet is very large it may be necessary to widen the gap slightly. An adjustment is provided on a beet harvester to allow for this setting and it may take the form of using packing washers between the share arms and the side of the harvester.

Alternative types of shares are available to suit adverse conditions such as where the soil may be exceptionally wet and heavy. These shares are designed to remove some of the soil that may be adhering to the beet when it is lifted.

Oppel Wheels

These are an alternative lifting device to the lifting shares and they take the form of a pair of fairly large-diameter metal wheels angled and set in such a way that they squeeze the beet out of the ground as they travel along the row. They are positioned approximately where the normal type of lifting share would be and they rotate as the machine moves forward. The fact that they rotate whilst working makes it easier for the tractor to pull the harvester and this can be an advantage when conditions are difficult. Figure 170 shows an illustration of this mechanism which requires a flail device positioned behind it to flick the beet on to the elevator. Depth setting for Oppel wheels is similar to that required for lifting shares.

Cleaning and Elevating the Beet

There is always some soil adhering to some of the beet after it is removed from the ground. How much will depend on such factors as type of soil in which the crop is grown, weather conditions, setting of the lifting shares, etc., but it is only in exceptionally dry seasons that the amount of soil lifted with the beet is to a minimum.

Fig. 170. Principle of operation of the Oppel Wheel beet-lifting mechanism.

Because more often than not the beet is lifted with varying amounts of soil adhering to it, it is necessary to have on the beet harvester itself a device that will assist in cleaning the beet. All harvesters are, therefore, so equipped. The method, of course, varies from one make of harvester to another but a typical method is to make use of agitating devices which cause the elevating web to "bounce" thus causing the beet to be thrown about on its way up the elevator. The elevating web of all beet harvesters is of the open type with straight or waved links so that skuffing of the beet takes place and any soil knocked off the beet can fall through the web and back on to the field.

Another typical method of cleaning beet on the harvester is to have a stationary web hanging above the moving elevator web and in such a position that the beets are skuffed between the two. This has the effect of rolling the beet as it moves up the elevator, and all sides of it are scrubbed by the elevator links. A more drastic cleaning effect can be achieved by hanging weights to this stationary web thus slowing down the movement of the beet as well as increasing the scrubbing action.

From this elevator the beet is deposited on to a cross elevator which in turn elevates the beet into a trailer travelling alongside the harvester, unless of course the beet harvester is a tanker model in which case the beet will be elevated into the tank.

Multi-row and Self-propelled Beet Harvesters

Numerous types of beet harvesters are available ranging from single-row tractor-hauled power-driven types, to self-propelled single- or double-row types. The former may be fitted with a self unloading tank of between about 1500-2500 kg (30-45 cwt approx.) capacity, or equipped to elevate straight into a trailer. The self-propelled types are all fitted with tanks of varying capacities ranging from 1000 to 3500 kg (20-70 cwt approx.). These latter machines can work at a fairly high output but one of their limitations is often made evident during difficult and wet seasons. This limitation is produced by their own weight coupled with the weight of beet being carried in the tank so that if field conditions are wet travelling can be very difficult if not impossible.

In addition to complete harvesters mentioned above, there is also a range of machines available which may be used to harvest sugar beet in more than one operation. For example, one machine may be used to top five or six rows of beet, to be followed by another machine which lifts and windrows the five or six rows into a single windrow. A third machine follows and elevates the windrow into a trailer. Although this system involves three separate machines in different operations the output is high and where large acreages of beet are grown may well be the best system to use.

An alternative type of harvester is the "Armer" machine (Fig. 171) which is a complete single-stage harvester but the sequence of events is somewhat different to that of conventional machines.

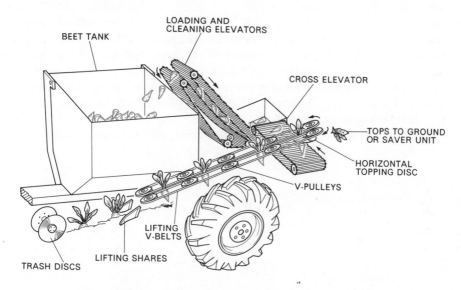

Fig. 171. A view of some of the main working parts of a complete
single row beet harvester.

The beet is lifted from the ground by means of two contra-rotating rubber belts which grip the beet by its top and as the machine progresses forward the beet are pulled out of the soil, assisted by a small share running under the crop. As the beet reach the top of the rubber belt the tops are removed by revolving blades which cause the "topped" beet to fall onto the cleaning web whilst the tops are carried over the rear of the machine and fall onto the ground in a windrow. Clean beet is elevated into a hopper for eventual discharge into a trailer at the headland.

These machines are trailed and powered by the p.t.o. and are available as single- or two-row units. Sensing arms on each side of the row automatically keep the lifting belts in line with the sugar beet by means of an hydraulic ram which moves the machine from side to side to ensure correct positioning.

Routine Maintenance of the Beet Harvester

The beet harvester can be described as a rugged machine requiring not a lot of servicing considering the work that it does.

However, a number of points will require daily attention whilst the machine is in use, these are as follows:

(a) Lubrication—carry out as recommended by the manufacturers, there are gear-boxes, greasing points and oiling points on most beet harvesters.

(b) Keep the topping knife sharp. Carry a spare if possible.

(c) Keep drive chains correctly tensioned and in correct alignment with sprockets.

(d) Keep V-belt drives correctly tensioned.

(e) Slip clutches, which are a protection device for the machine, should be properly adjusted.

(f) Keep all guards in order and securely positioned.

When the harvest is finished, the harvester should be thoroughly cleaned down to remove all soil, beet tops and rubbish. V-belt drives should be released of tension, slip clutches should have the spring pressure released and oil poured between the clutch plates. Alternatively, the plates can be taken off, cleaned, oiled and put back in position but without spring pressure applied. Drive chains should be removed, cleaned, soaked in oil and stored in an oiled condition either back in position on the machine or in the farm store or workshop. A coat of anti-rust preparation on the knife will help to prevent the knife edge rusting away.

All grease points should be lubricated particularly if the machine has to stand outside in all weathers.

CHAPTER 28

The Electric Fencer

ON MOST farms where livestock are kept, the electric fence forms part of the standard equipment used in the management and control of stock. It is widely used on dairy and beef farms where open and strip grazing is practised and for the control of pigs kept on open range. Sheep, horses and goats can also be effectively fenced in this manner. Not only is the electric fence useful for keeping livestock within a fenced area, but it may also be used to keep certain animals out. Foxes and dogs, for example, may be prevented from attacking poultry and sheep.

When the electric fence is in operation, any animal that touches the fencing wire receives an electric shock. Although the shock is not great enough to harm the animal, it is sufficient to make it avoid the fence again if possible.

Two main types of equipment are used for the operation of these fences, one being supplied by mains electricity and the other by a battery. The latter is more common and will be dealt with here.

To provide a complete working electric fence, the following items of equipment are needed. A 6-V battery, control unit, suitable fencing posts, insulators, fencing wire and wire strainer.

The Battery

Batteries used for fencers are 6 V and may be one of two types: one of these is a dry battery which is not rechargeable and may have a working life of about 4 months, and the other is a chargeable accumulator which requires recharging every 3 weeks or so if in constant use.

The Control Unit

The control unit, Fig. 172, is the heart of the fencing system because it provides the "shock" and it is designed to do this approximately 60 times per minute. It is made up of a magnetic core, two coils, an armature, contact points, a flywheel and spiral spring, and suitable leads to battery and fence.

The core is made of soft iron strips clamped together and forming a U shape. Bridged across the ends of the U is a metal armature which is actually fixed to one leg of the U only. The other end does not touch the other leg of the core but it extends past it and is fitted with a contact point. This contact point rests on another fitted to a flywheel which is free to rotate on a spindle. A spiral spring is fixed to the spindle so that it will wind or unwind if the flywheel rotates.

Two coils are wound on the iron core, one being a primary coil, and the other a secondary coil, their wiring arrangement being shown. The battery and earth lead complete the outfit.

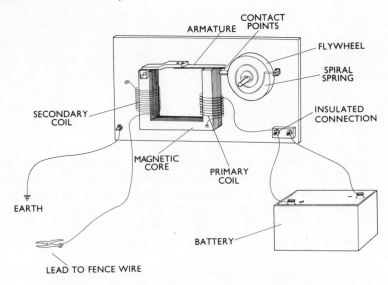

Fig. 172. Principle of the operation of a battery-operated electric fencer.

How It Works

When current passes through the primary coil, the iron core is magnetized and this causes the armature to be attracted smartly downwards closing the gap between itself and the leg of the iron core. This movement pushes the contact point on the flywheel and starts the flywheel rotating. As soon as the points separate, the primary circuit is broken and the magnetic field collapses allowing the armature to spring back to its original position. Also a high-voltage current is induced in the secondary coil which is connected to the fence wire, so that any animal touching the fence at that moment is given a sharp shock.

When the flywheel rotates it causes the spiral spring to unwind, but this eventually stops the rotation and reverses it so that the wheel swings back and the contact points come together again. When this happens, the primary circuit is again complete, the iron core is again magnetized and the cycle of operation is repeated. The cycle is repeated approximately 60 times per minute and continues so long as the battery is connected and is in a charged condition.

An indication that the battery is in a low state of charge is when the control unit is "ticking" at a faster rate than normal. This ticking, which can be heard when the fencer is working, is caused by the points striking when the flywheel is operating. A faster rate of "tick" is due to a reduced magnetic effect which in turn is caused by a low battery voltage. The reduced magnetic effect results in a less severe push to the flywheel by the armature, so that the flywheel does not make its complete swing. It therefore returns quicker.

The Fence

For satisfactory and trouble-free operation the fence must be properly erected and to do this it is necessary to have suitable end posts (strainers), intermediate posts, insulators

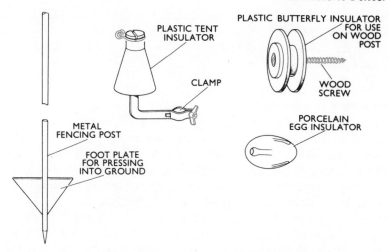

Fig. 173. Types of insulators and a fencing stake for use on an electric
fence.

and fencing wire. The wire must be kept taut and this can be done by the use of a
tensioning device.

The posts used may be of metal or wood but whichever is used it is necessary to have
an insulator fitted to each one to carry the wire. Generally the posts may be spaced
about 13·5 m (15 yd approx.) apart. Many different types of insulators are available
for fitting to posts of wood or metal, a common type used on metal posts being known
as a tent insulator, whilst a type used on wooden posts is a reel or butterfly insulator.
Both are made of plastic material. For use at corners an egg-type insulator may be
used and these are made of porcelain. Figure 173 shows these types of insulators and
a typical metal post to which the insulator can be clamped at any height.

Figure 174 shows two fence arrangements, one having a single wire strand and the
other a double strand. The type of wire used for the fence is generally 2 mm (14 gauge)

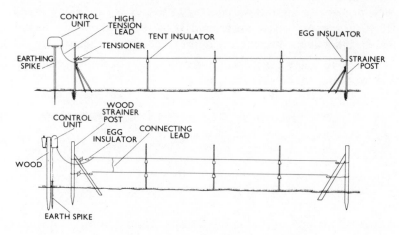

Fig. 174. Two methods of erecting an electric fence.

galvanized wire, but if necessary barbed wire may be used. This may be found to be more effective when fencing sheep.

To electrify the double strand fence it is only necessary to connect these strands together at one point.

Causes of Faulty Working

As soon as the high-tension lead is clipped on to the fencing wire and the battery is connected and switched on, the electrical impulses created by the control unit charge the fence wire. Because the wire is supported on insulators, the current will not flow from it unless some object in contact with the ground touches the wire and completes the circuit. Should an animal or human being touch the wire they will complete the circuit, and at the same time receive an uncomfortable shock.

If the grass or weeds are allowed to touch the wire, they also will complete the circuit with the result of a greatly reduced intensity of shock to any animal that may touch the wire at the same time. This may lead to animals breaking out because it is not the fence itself that prevents them but the fear of getting a shock. A similar result would occur if a faulty insulator was used, thereby allowing the current to go to earth through the fencing post.

For efficient operation it is also necessary to have a good earth connection to the fencer unit. Generally, if the unit is attached to a metal mounting post which is driven into the ground, a separate earth connection is not required. However, if a wooden mounting post is used it will be necessary to drive a metal stake well into the ground and connect it to the fencer unit with suitable wire.

Height of Fence

The height at which the fencing wire is positioned depends largely on the size and type of animal being fenced in. Generally, the following heights are recommended:

(a) *Cattle*: one strand of wire 76 cm (2 ft 6 in. approx.) from the ground.
(b) *Sheep:* two strands of wire, one 30 cm (12 in. approx.) and the other about 60 cm (24 in. approx.) from the ground.
(c) *Pigs*: one or two strands—according to size of pigs. If one strand, 20-30 cm (10-15 in. approx.) above the ground, if two strands, one 25 cm (10 in. approx.) and the other about 40 cm (16 in. approx.) above the ground.

A general rule to follow regarding setting the height of the fence is to set it between a half and two-thirds the height of the animal.

CHAPTER 29

Milking Machines

THE need for efficient handling of milk on the scale as practised in Great Britain has led to considerable developments in milking machines, particularly in terms of bulk handling. Although the basic milk-extraction process has remained unchanged for many years the necessity to do away with the handling of churns and bucket units has given rise to the fitting of pipeline milking systems throughout all major dairy farms.

The layout of a pipeline parlour complete with automatic washing is shown in Fig. 175.

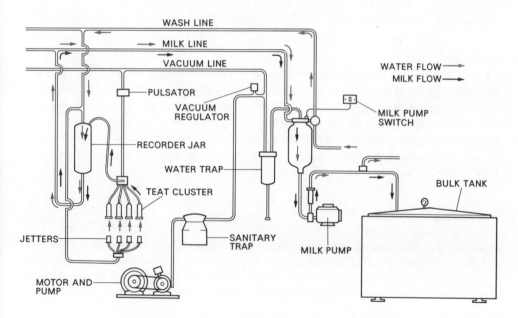

Fig. 175. A modern milking parlour layout including a wash system.

The Vacuum Pump

These are normally of the sliding vein type (Fig. 176) and are driven by electric motor although it is desirable that an emergency power supply is available should the electric supply fail. This may be achieved by providing either a tractor p.t.o. drive to the pump or a hydraulic motor which may also be operated by the tractor.

Vacuum pumps require little maintenance and periodic checking of drive belt tension along with topping-up of the oil reservoir is usually sufficient.

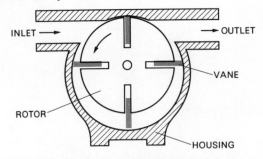

INLET →   → OUTLET

VANE

ROTOR

HOUSING

Fig. 176. A vacuum pump.

The pump draws air from the milking system and should have sufficient capacity to be able to maintain a vacuum of 380 mm of mercury throughout the milking period. This level of vacuum should be sustained even if a unit is kicked off by a cow or as units are fitted or removed from an animal, consequently the capacity of the pump needs to be well in excess of normal requirements. The minimum capacity can be expressed as $150 + 60N$ litres/minute, where N = number of units.

Excess pump capacity is controlled by the vacuum regulator.

The Sanitary Trap

As a precaution against dirt or moisture entering the pump, a sanitary trap is fitted in the pipeline. This then traps the dirt and moisture should any be sucked through the pipe. It is usually necessary to attend to the sanitary trap weekly, in some cases more often, to remove moisture collected in the pail. This moisture may get there due to condensation in the pipeline or when the pipeline is washed out with water. Whenever the pail is removed, it is important that it is always firmly secured in position again in order to prevent leakage which will cause a loss of vacuum.

Vacuum Regulator (Fig. 175)

Situated in the vacuum line this controls the level of vacuum within the system. It is important that 380 mm of mercury vacuum is maintained without fluctuations which may have a harmful effect on the cow. The regulator is continually opening and closing during the milking process and admitting air into the system when need be to compensate for excessive pump capacity when no natural air admission is taking place.

Vacuum Gauge

So that the operator will know the amount of vacuum being produced when the plant is in operation a vacuum gauge is fitted to the pipeline. This gauge will generally be calibrated in millimetres of mercury and a safe working level is between 330 and 380 mm (13-15 in.). This is approximately half atmospheric pressure.

Water Trap

The water trap is installed to separate the air (vacuum) side of the milking system from the liquid (milk and washing water) side. It comprises of a glass jar, so that the operator can detect any defect, fitted with a float valve. This valve will automatically shut off the vacuum pipe should liquid enter the jar, thus protecting the pump from possible damage. The jar is also fitted with a self-closing foot valve to permit liquid to drain out when the plant is not working. From the water trap the vacuum is transferred to the intercepter vessel from where the vacuum is extended into the milk line and the wash line.

Principle of milk extraction:

The teat cup assembly comprises of a rigid outer shell made of stainless steel into which is fitted a synthetic rubber liner which is flexible and therefore free to distort. A constant vacuum is exerted in the long milk tube, through the claw, via the short milk tube into the centre of the liner, thus drawing milk from the animal. Connected via the pulsator is an intermittent vacuum supply to the area surrounding the liner, so that when vacuum is applied to this area the pressure on the inside and outside of the liner is equal, the liner retains its natural shape and milk flows. When air is admitted to the outside of the liner we now have a situation of low pressure inside, high pressure (atmospheric) outside, this causes the liner to collapse, squeezing the teat and at the same time shutting off the vacuum sucking at the teat and consequently milk flow stops (Fig. 177).

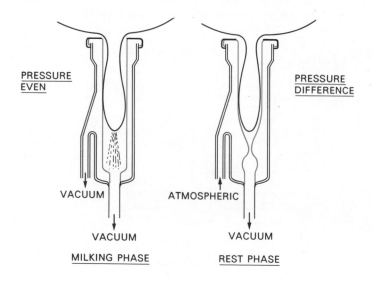

PRESSURE EVEN — PRESSURE DIFFERENCE

VACUUM — ATMOSPHERIC

VACUUM — VACUUM

MILKING PHASE — REST PHASE

Fig. 177.

This alternate supply of vacuum/air is known as the PULSATION RATE and occurs approximately 60 times/min. The relationship between vacuum/air or MILK:REST is called the PULSATION RATIO which is normally 3:1.

Milk flows from the teat cup to the claw from where it travels via the long milk tube to the recorder jar. Here the milk is held until the cow has finished milking, permitting recording, sampling or even rejection if need be, before being transferred through the milk line to the interceptor vessel.

As the milk level rises in this vessel it eventually switches on the milk pump which conveys the milk out of the milking system which is under vacuum and into the bulk milk tank (atmospheric) where the milk is cooled and stored to await collection by tanker.

Automatic Washing

To facilitate the cleaning of a pipeline system there are two methods of automatic washing:

1. Circulation cleaning.
2. Acidified boiling water cleaning.

The two systems vary slightly in their method but both operate by drawing in the cleaning liquid into the system from where it is carried by the wash line to all parts of the milking plant, including the jetters which have been placed on to the teat cups to facilitate the cleaning of the cups along with the short and long milk tubes.

Bulk Storage

To prevent the rapid growth of bacteria within the milk it is essential that it is stored under refrigerated conditions. Figure 178 shows a typical bulk tank capable of cooling milk down to the required storage temperature of 4·4°C (40°F).

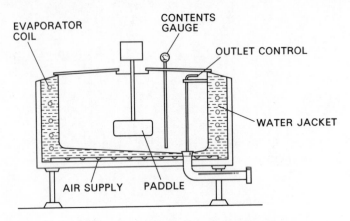

Fig. 178. A section through a bulk milk tank.

The milk within the tank imparts its heat to the water in the surrounding jacket which is chilled by the formation of an ice bank created by the refrigeration unit. On some models of bulk tank air is bubbled through the chilled water to improve the cooling. Inside the tank are paddles which stir the milk to ensure an even temperature throughout the entire mass.

Refrigeration

The principle of refrigeration is based on the fact that if a liquid is pressurized, the temperature at which it boils will be raised. Figure 178 shows the basic circuit consisting of a low-pressure evaporation coil, a compressor, and a high-pressure condensing coil. The chemical within the circuit turns from a liquid to gaseous state whilst in the evaporation coils and in doing so draws heat from the surrounding mass, so it is these coils which are situated in the bulk tank jacket. The gas is then compressed by the compressor and in doing so reverts back to a liquid whilst at the same time giving off its heat to the condensing coils which are usually situated outside the dairy and cooled by the surrounding air.

CHAPTER 30

Safety on the Farm

A BOOK dealing with the subject of farm machinery cannot be complete without including some mention of farm safety. This is particularly so in this age of increasing mechanization on farms when there has also been an increase in the number of hazards facing the farm machinery operator and farm worker.

Each year there are many accidents on or about farms when people are killed or injured. Many of these accidents are caused by carelessness and ignorance and many are caused by neglect. It is not uncommon for a tractor driver to be crushed to death by his tractor overturning on top of him, nor is it uncommon for a farm worker to die because of poisoning due to careless handling of spray chemicals. There are numerous other examples of how people are killed or injured on farms.

A few are listed below:

Falling off trailers and under wheels.
Electrocution due to faulty equipment.
Buried in grain and suffocating after falling into a grain bin.
Dismounting from a tractor whilst it is moving and falling under a wheel.
Shot by own gun after falling whilst climbing a fence.

These are but a few causes of death on farms, but in addition to these many limbs are broken or lost and a common cause of such accidents is when machinery operators attempt to adjust their machines whilst they are still operating. This is extremely foolish practice.

In an attempt to reduce the number of accidents on farms many regulations have been introduced governing the safety requirements for farm machinery and farm workplaces. These regulations lay down certain responsibilities to be undertaken by employers and employees on farms and they must be complied with. Failure to do so can result in a fine.

The problem of introducing safety regulations to cover the use of such a wide variety of machinery and equipment used on farms is not a small one and has been done in stages. At the time of writing, some regulations are now in force and some are arranged to come into force at different stages during the years to come. It is not within the scope of this book to give detailed information or interpretation of all the regulations existing now or that are to be introduced in the future, but to give the reader knowledge of some of the requirements of the regulations and causes of accidents. He will then be aware of what is being done and what he should do for his and other people's safety.

The various Acts and Regulations can be obtained from H.M. Stationery Offices and booksellers and there are many excellent leaflets to be obtained free of charge from Offices of the Ministry of Agriculture. Safety inspectors in the employ of the Ministry of Agriculture are always pleased to give assistance and advice.

Safety when using Tractors

Many of the accidents involving tractors need never happen if drivers avoid taking risks and also see that their tractors are kept mechanically sound.

Few drivers realize how easy it is to overturn a tractor. In Chapter 9, this is discussed in the paragraphs dealing with centre of gravity, and stability of tractors and machines. It is very important.

A tractor will overturn quite easily rearwards so that it crushes the driver directly beneath it. This is invariably caused by towing a machine or heavy load from a high hitch point, usually the top link attachment point, at the rear of the tractor. The risk is greatly increased if the load is also being towed up an incline, and the steeper the incline the greater the risk. It is always wise to tow a load from the proper drawbar only.

A tractor must be mechanically sound to be safe. Apart from the engine, the mechanisms in the transmission system must be in good order. An awkward situation can develop if due to wear in the gearbox the gears disengage on their own when towing a load up or down an incline. Steering mechanisms on a tractor must be sound and positive. Brakes must be properly adjusted and efficient. This is particularly important and checking of the adjustment and balancing of the brakes should be done frequently.

P.t.o. shafts are a hazard and many people have been severely injured through getting their clothing caught in them. There is now a regulation which makes it an offence to operate such a shaft without it being properly guarded. This applies to the short end of the p.t.o. protruding from the rear of the tractor which must be guarded at all times when the engine is running, and to the p.t.o. shaft extending to the machine being driven.

It is the employer's responsibility to see that these guards are fitted and workers must not use any tractor or machine when they are not fitted.

Other regulations concerning the use of tractors indicate that it is unlawful for children under the age of 13 years to ride on tractors, self-propelled machines or any other machine or implement carried or towed by a tractor. Two persons are not allowed to ride on a tractor if it is not equipped with seating for two. Nor is anyone allowed to ride on the drawbar of a tractor or machine.

Safety when using Field Machinery

The wide variety of machines used on farms makes possible many different ways in which accidents can take place. To impress on the reader the seriousness of such accidents it is necessary to give more examples of just what does happen.

Cutter-bars on mowing machines, binders and combine harvesters usually claim victims each year in the form of amputated fingers. The writer recalls one such victim who was cutting a hay crop with a mowing machine. The mowing machine was power driven to a pulley which drove the knife through double V-belts. Correct tension on the belts allowed the knife to stop reciprocating when an obstruction wedged between the knife and cutter-bar fingers. The belts slip and the knife and fingers are protected against damage. In this particular instance a stone wedged between a knife section and a cutter-bar finger thus allowing the belt to slip and preventing the knife from moving. On seeing the knife stop moving the driver dismounted from his tractor to see what was wrong. He saw the stone obstructing the knife and immediately removed it with his finger whereupon the knife started moving again and cut off his finger. He had made the silly mistake of not putting the power drive out of gear before getting off his tractor to attend to the obstruction.

A similar accident known to the writer involved a combine harvester driver. He was cleaning out the grain tank of his combine through the small cleaning door at the bottom of the tank. Whilst he had his hand inside removing the last few grains, another worker helping him to clean out the combine, started up the engine. The grain-unloading auger in the tank happened to have been left in gear and it rotated and cut off the driver's thumb.

The reader may not think that these are serious accidents, but he can be assured that more incredible things do happen on farms. Such as a tractor driver passing through the shredding mechanisms of a power-driven manure spreader or another driver being beheaded when he released the oil out of the lift cylinders of an uplifted fore-end loader so that it immediately dropped. Or of the driver who slipped and was pulled beneath the spinning wheel of a tractor and found hours later with the wheel still spinning on top of him.

These are silly accidents that need never have occurred if care had been taken, yet every day similar accidents happen. It is not uncommon for a worker on a grain drill to fall backwards off the drill platform and in front of the harrows being dragged behind the drill. Many workers are injured by getting clothing and limbs caught in drive belts, chains and pulleys.

Many regulations are now in force concerning the guarding and use of field machinery and again these regulations must be adhered to. The regulations in force now cover a number of machines now in use on farms, but eventually all field machinery will be within their scope. All new machinery sold to farmers has to be fitted with guards according to the requirements.

The following examples will give the reader some idea of what has been done on some machines to make them safer to use.

Open shafting, sprockets, chains, pulleys, belts and gears must be properly guarded so as to prevent a worker or machinery operator coming in contact with them. Grain drills must be fitted with a hand rail for use by the drill man so that he does not fall off the drill. They must also have a toeboard along the front edge of the platform to prevent the drill man slipping forward into the working parts of the drill. Machines fitted with cutter-bars must have a rigid guard over the fingers of the bar when the cutter-bar is *not* in use. This is necessary because anyone walking into the points on the fingers of a cutter-bar can get a nasty injury.

Safety around the Farmstead

There are further regulations applicable to safe working conditions in and around the farmstead. These cover stairs, ladders, grain pits, grain bins, etc. Stairs must be in a good state of repair and have no steps missing. They must be fitted with hand-rails if the stairs are more than 0·914 m (3 ft) high. Ladders must be in good condition, no rungs missing or damaged. It is necessary to put guard rails around grain pits or bins if there is risk of a worker falling more than 1·524 m (5 ft) in depth.

There are regulations covering the safety requirements for stationary machines. Such machines as grain driers, dressers, hammer mills, mixers, elevators and stationary engines are in this category. As with field machinery it is necessary to guard belts, shafts, pulleys, chains and sprockets, etc., but with grain handling and processing machinery it is also a requirement to guard the inlet points where the grain is taken in, say by an auger, and at the outlet where it is discharged. This must be done to prevent a worker coming in contact with moving parts, and it also applies to mills, mixers, root cutters, etc.

Safety when Handling Spray Chemicals

There can be many hazards associated with using spray chemicals, not only to the user but to other people, animals, wild-life, fish, and crops. A user should therefore take care when handling the chemicals and spraying.

Certain precautions are required by law when some types of chemicals are used and the user should make sure that he knows what these precautions are. Information about this is given in leaflets issued by the Ministry of Agriculture. If some chemicals are used that do not come within the scope of the law with regard to handling and spraying, it is still wise to be careful and take some precautions. These precautions can be listed as follows:

(a) Make sure that you read any instructions given by the manufacturer of the chemical.

(b) If it is recommended that protective clothing be worn, then make quite sure that you do so. If nothing specific is laid down with regard to protective clothing, it is still wise to wear rubber gloves when handling the chemical and a face mask when spraying.

(c) The body can be contaminated by the chemical in many ways:
   (i) when spraying, due to the drift contacting the skin;
   (ii) when handling the chemical without suitable protection of the skin;
   (iii) when smoking whilst working, and the spray drift gets on to the cigarette and lips;
   (iv) when eating because of contamination of the hands and mouth due to spray drift;
   (v) from clothing which is contaminated with spray chemicals.
   Because of this, sprayer operators must take a few simple precautions to avoid the contamination. The wearing of suitable clothing has been mentioned. He must wash hands and face thoroughly before eating and drinking. He should avoid smoking when working. He should try to keep out of the spray drift if possible. This means do not spray if it is too windy but if there is some breeze spray in a direction across the breeze.

(d) Spray drift can do damage to other crops that are susceptible to the chemical being used, therefore the direction of the wind must be known in case serious damage is done to neighbouring crops in fields and in gardens.

(e) Avoid spray drift going on to grazing pastures where animals may graze the contaminated pasture.

(f) When spraying is finished, care must be taken in the disposal of any liquid left over and the water used for washing down the spraying machine. If this is discharged where it can drain into rivers, streams or ponds, fish may be killed also wild birds and animals if their drinking places are polluted.

(g) Containers that have been used for spray chemical should be suitably disposed of and not put where children and animals can get at them.

First Aid on the Farm

Whatever the precautions taken when working with machinery and equipment on farms, some accidents will still take place, and if they do, first aid attention will be necessary. A first aid box is therefore essential. In fact by law a first aid box must be kept on a farm. The number of dressings and their sizes, etc., are specified in the First Aid Regulations which can be obtained from H.M. Stationery Office and booksellers.

CHAPTER 31

Metric Tables

Length

10 mm	= 1 cm	(mm	= millimetre)
10 cm	= 1 dm	(cm	= centimetre)
10 dm	= 1 m	(dm	= decimetre)
10 m	= 1 dam	(m	= metre)
10 dam	= 1 hm	(dam	= decametre)
10 hm	= 1 km	(hm	= hectometre)/(km = kilometre)

Area

100 mm²	= 1 cm²	(mm²	= square millimetres)
100 cm²	= 1 dm²	(cm²	= square centimetres)
100 dm²	= 1 m²	(dm²	= square decimetres)
100 m²	= 1 Ar	(m²	= square metres)
100 Ar	= 1 har	(Ar	= Are)
100 har	= 1 km²	(har	= hectare)/(km² = square kilometre)

Volume

1000 mm³	= 1 cm³	(mm³	= cubic millimetres)
1000 cm³	= 1 dm³	(cm³	= cubic centimetres)
1000 dm³	= 1 m³	(dm³	= cubic decimetres)/(m³ = cubic metres)

Weight

10 mg	= 1 cg	(mg	= milligramme)
10 cg	= 1 dg	(cg	= centigramme)
10 dg	= 1 g	(dg	= decigramme)
10 g	= 1 dag	(g	= gramme)
10 dag	= 1 hg	(dag	= decagramme)
10 hg	= 1 kg	(hg	= hectogramme)/(kg = kilogramme)

Dry and Liquid Measure

10 ml	= 1 cl	(ml	= millilitres)
10 cl	= 1 dl	(cl	= centilitres)
10 dl	= 1 l	(dl	= decilitres)
10 l	= 1 dal	(l	= litre)
10 dal	= 1 hl	(dal	= decalitre)
10 hl	= 1 kl	(hl	= hectolitre)/(kl = kilolitre)

Conversion Table

25·4 mm	= 1 inch	0·039 in.	= 1 mm
2·54 cm		0·394 in.	= 1 cm
30·48 cm	= 1 foot	39·371 in.	
0·914 m	= 1 yard	3·28 ft	= 1 m
20·1 m	= 1 chain	1·094 yd	
1609·33 m	= 1 mile	0·621 miles	= 1 km

6·451 cm²	= 1 in.²	0·155 in.²	= 1 cm²
929 cm²	= 1 ft²	10·76 ft²	= 1 m²
836 cm²	= 1 yd²	1·196 yd²	= 1 m²
4046·7 m²	= 1 acre	0·3861 mile²	= 1 km²
2·59 km²	= 1 mile²		

16·387 cm³	= 1 in.³	0·061 in.³	= 1 cm³
28·316 dm³	= 1 ft³	61·027 in.³	= 1 litre
76·453 dm³	= 1 yd³	0·22 gal	= 1 litre
1 litre	= 1·759 pints		
1 kg	= 2·2 lb	1 lb	= 0·4536 kg
1 Metric ton	= 2204·6 lb	1 ton	= 1016 kg
			= 1·016 metric tons

To convert	*to*	*Multiply by*
inches	centimetres	2·54
feet	metres	0·3048
miles	kilometres	1·609
square inches	square centimetres	6·452
square feet	square metres	0·0929
acres	square metres	4046·85
pints	litres	0·5682
gallons	litres	4·5459
pounds	kilogrammes	0·4535
cwts	kilogrammes	50·8023
tons	kilogrammes	1016·05
lb per sq. in.	kilos per sq. cm	0·07031
tons per acre	kilos per hectare	2510·71
ft-lb	kilogramme metres	0·1383
in.-lb	kilogramme metres	0·0115

Index